# NJU SA 2013-2014

THE YEAR BOOK OF ARCHITECTURE PROGRAM SCHOOL OF ARCHITECTURE AND URBAN PLANNING
南京大学建筑与城市规划学院建筑系　教学年鉴
王丹丹，华晓宁 编　EDITORS: WANG DANDAN, HUA XIAONING
东南大学出版社·南京　SOUTHEAST UNIVERSITY PRESS, NANJING

### 建筑设计及其理论
### Architectural Design and Theory

| | |
|---|---|
| 张　雷 教　授 | Professor ZHANG Lei |
| 冯金龙 教　授 | Professor FENG Jinlong |
| 吉国华 教　授 | Professor JI Guohua |
| 周　凌 教　授 | Professor ZHOU Ling |
| 傅　筱 副教授 | Associate Professor FU Xiao |
| 钟华颖 讲　师 | Lecturer ZHONG Huaying |

### 城市设计及其理论
### Urban Design and Theory

| | |
|---|---|
| 丁沃沃 教　授 | Professor DING Wowo |
| 鲁安东 教　授 | Professor LU Andong |
| 华晓宁 副教授 | Associate Professor HUA Xiaoning |
| 刘　铨 讲　师 | Lecturer LIU Quan |
| 胡友培 讲　师 | Lecturer HU Youpei |
| 尹　航 讲　师 | Lecturer YIN Hang |

### 建筑历史与理论及历史建筑保护
### Architectural History and Theory, Protection of Historic Building

| | |
|---|---|
| 赵　辰 教　授 | Professor ZHAO Chen |
| 王骏阳 教　授 | Professor WANG Junyang |
| 肖红颜 副教授 | Associate Professor XIAO Hongyan |
| 胡　恒 副教授 | Associate Professor HU Heng |
| 冷　天 讲　师 | Lecturer LENG Tian |

### 建筑技术科学
### Building Technology Science

| | |
|---|---|
| 鲍家声 教　授 | Professor BAO Jiasheng |
| 秦孟昊 教　授 | Professor QIN Menghao |
| 吴　蔚 副教授 | Associate Professor WU Wei |
| 郜　志 副教授 | Associate Professor GAO Zhi |
| 童滋雨 副教授 | Associate Professor TONG Ziyu |

南京大学建筑与城市规划学院建筑系
Department of Architecture
School of Architecture and Urban Planning
Nanjing University
arch@nju.edu.cn　http://arch.nju.edu.cn

教学纲要
# EDUCATIONAL PROGRAM

| 教学阶段 Phases of Education | 本科生培养（学士学位）Undergraduate Program (Bachelor Degree) | | | |
|---|---|---|---|---|
| | 一年级 1st Year | 二年级 2nd Year | 三年级 3rd Year | 四年级 4th Year |
| 教学类型 Types of Education | 通识教育 General Education | | 专业教育 Professional Train | |
| 课程类型 Types of Courses | 通识类课程 General Courses | 学科类课程 Disciplinary Courses | | 专业类课程 Professional Courses |
| 主干课程 Design Courses | 设计基础 Basic Design | 建筑设计基础 Basic of Architectural Design | 建筑设计 Architectural Design | |
| 理论课程 Theoretical Courses | 专业基础理论 Basic Theory of Architecture | 专业理论 Architectural Theory | | |
| 技术课程 Technological Courses | | | | |
| 实践课程 Practical Courses | 环境认知 Environmental Cognition | 古建筑测绘 Ancient Building Survey and Drawing | 工地实习 Practice of Construction Plant | |

| 研究生培养（硕士学位）Graduate Program (Master Degree) | | | 研究生培养（博士学位）Ph. D. Program |
|---|---|---|---|
| 一年级 1st Year | 二年级 2nd Year | 三年级 3rd Year | |

学术研究训练 Academic Research Training

学术研究 Academic Research

| 建筑设计研究 Research of Architectural Design | 毕业设计 Thesis Project | 学位论文 Dissertation | 学位论文 Dissertation |
|---|---|---|---|
| 专业核心理论 Core Theory of Architecture | 专业扩展理论 Architectural Theory Extended | 专业提升理论 Architectural Theory Upgraded | 跨学科理论 Interdisciplinary Theory |

建筑构造实验室 Tectonic Lab

建筑物理实验室 Building Physics Lab

数字建筑实验室 CAAD Lab

生产实习 Practice of Profession　　生产实习 Practice of Profession

# 课程安排
# CURRICULUM OUTLINE

| | 本科一年级<br>Undergraduate Program 1st Year | 本科二年级<br>Undergraduate Program 2nd Year | 本科三年级<br>Undergraduate Program 3rd Year |
|---|---|---|---|
| 设计课程<br>Design Courses | 设计基础<br>Basic Design | 建筑设计基础<br>Basic Design of Architecture<br>建筑设计（一）<br>Architectural Design 1 | 建筑设计（二）<br>Architectural Design 2<br>建筑设计（三）<br>Architectural Design 3<br>建筑设计（四）<br>Architectural Design 4<br>建筑设计（五）<br>Architectural Design 5 |
| 专业理论<br>Architectural Theory | 逻辑学<br>Logic | 建筑导论<br>Introductory Guide to Architecture | 建筑设计基础原理<br>Basic Theory of Architectural Design<br>居住建筑设计与居住区规划原理<br>Theory of Housing Design and Residential Planning<br>城市规划原理<br>Theory of Urban Planning |
| 建筑技术<br>Architectural Technology | 理论、材料与结构力学<br>Theoretical, Material & Structural Statics<br>Visual BASIC程序设计<br>Visual BASIC Programming | CAAD理论与实践<br>Theory and Practice of CAAD | 建筑技术（一） 结构与构造<br>Architectural Technology 1: Structure & Construction<br>建筑技术（二） 建筑物理<br>Architectural Technology 2: Building Physics<br>建筑技术（三） 建筑设备<br>Architectural Technology 3: Building Equipment |
| 历史理论<br>History Theory | 古代汉语<br>Ancient Chinese | 外国建筑史（古代）<br>History of World Architecture (Ancient)<br>中国建筑史（古代）<br>History of Chinese Architecture (Ancient) | 外国建筑史（当代）<br>History of World Architecture (Modern)<br>中国建筑史（近现代）<br>History of Chinese Architecture (Modern) |
| 实践课程<br>Practical Courses | | 古建筑测绘<br>Ancient Building Survey and Drawing | 工地实习<br>Practice of Construction Plant |
| 通识类课程<br>General Courses | 数学<br>Mathematics<br>语文<br>Chinese<br>名师导学<br>Guide to Study by Famed Professors<br>计算机基础<br>Basic Computer Science | 社会学概论<br>Introduction of Sociology<br>社会调查方法<br>Methods for Social Investigation | |
| 选修课程<br>Elective Courses | | 城市道路与交通规划<br>Planning of Urban Road and Traffic<br>环境科学概论<br>Introduction of Environmental Science<br>人文科学研究方法<br>Research Method of the Social Science<br>美学原理<br>Theory of Aesthetics<br>管理学<br>Management<br>概率论与数理统计<br>Probability Theory and Mathematical Statistics<br>国学名著导读<br>Guide to Masterpieces of Chinese Ancient Civilization | 人文地理学<br>Human Geography<br>中国城市发展建设史<br>History of Chinese Urban Development<br>欧洲近现代文明史<br>Modern History of European Civilization<br>中国哲学史<br>History of Chinese Philosophy<br>宏观经济学<br>Macro Economics<br>管理信息系统<br>Management Operating System<br>城市社会学<br>Urban Sociology |

| 本科四年级 | 研究生一年级 | 研究生二、三年级 |
| --- | --- | --- |
| Undergraduate Program 4th Year | Graduate Program 1st Year | Graduate Program 2nd & 3rd Year |
| 建筑设计（六）<br>Architectural Design 6<br>建筑设计（七）<br>Architectural Design 7<br>本科毕业设计<br>Graduation Project | 建筑设计研究（一）<br>Design Studio 1<br>建筑设计研究（二）<br>Design Studio 2<br>数字建筑设计<br>Digital Architecture Design<br>联合教学设计工作坊<br>International Design Workshop | 专业硕士毕业设计<br>Thesis Project |
| 城市设计理论<br>Theory Urban Design | 城市形态研究<br>Study on Urban Morphology<br>现代建筑设计基础理论<br>Preliminaries in Modern Architectural Design<br>现代建筑设计方法论<br>Methodology of Modern Architectural Design<br>景观都市主义理论与方法<br>Theory and Methodology of Landscape Urbanism | |
| 建筑师业务基础知识<br>Introduction of Architects' Profession<br>建设工程项目管理<br>Management of Construction Project | 材料与建造<br>Materials and Construction<br>中国建构（木构）文化研究<br>Studies in Chinese Wooden Tectonic Culture<br>计算机辅助技术<br>Technology of CAAD<br>GIS基础与运用<br>Concepts and Application of GIS | |
| | 建筑理论研究<br>Study of Architectural Theory | |
| 生产实习（一）<br>Practice of Profession 1 | 生产实习（二）<br>Practice of Profession 2 | 建筑设计与实践<br>Architectural Design and Practice |
| 景观规划设计及其理论<br>Theory of Landscape Planning and Design<br>东西方园林<br>Eastern and Western Gardens<br>地理信息系统概论<br>Introduction of GIS<br>欧洲哲学史<br>History of European Philosophy<br>微观经济学<br>Micro Economics<br>政治学原理<br>Theory of Political Science<br>社会学定量研究方法<br>Quantitative Research Methods in Sociology | 建筑史研究<br>Studies in Architectural History<br>建筑节能与可持续发展<br>Energy Conservation & Sustainable Architecture<br>建筑体系整合<br>Advanced Building System Integration<br>规划理论与实践<br>Theory and Practice of Urban Planning<br>景观规划进展<br>Development of Landscape Planning | |

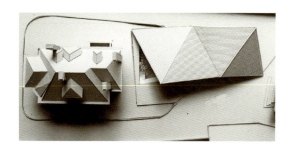

**1—13** 教学论文 ARTICLES ON EDUCATION

**15—117** 年度改进课程 WHAT'S NEW

**2**
关于"居住建筑设计原理"
TEACHING ON RESIDENTIAL BUILDING DESIGN PRINCIPLES

**16**
设计基础
BASIC DESIGN

**6**
以城市物质形态为基础的本科城市设计理论课程教学研究
A PEDAGOGICAL STUDY ON THE COURSE OF URBAN DESIGN THEORIES FOR UNDERGRADUATE STUDENTS, BASED ON URBAN PHYSICAL FORMS

**32**
建筑设计（一）风景区茶室设计
ARCHITECTURAL DESIGN 1  TEA HOUSE DESIGN

**10**
"建筑环境学"教学学生科研能力培养的探索
EXPLORATION INTO THE DEVELOPMENT OF STUDENTS' RESEARCH CAPABILITIES IN THE 'BUILDING ENVIRONMENT' COURSE

**40**
建筑设计（二）赛珍珠纪念馆扩建
ARCHITECTURAL DESIGN 2  EXPANSINO OF PEARL BUCK'S HOUSE

**48**
建筑设计（三）大学生活动中心设计
ARCHITECTURAL DESIGN 3  DESIGN OF THE COLLEGE STUDENT CENTER

**58**
建筑设计（四+五）社区商业中心+观演中心
ARCHITECTURAL DESIGN 4 & 5  COMMUNITY BUSSINESS CENTRE & PERFORMANCE CENTRE

**66**
本科毕业设计：洪家大屋测绘与改造设计
GRADUATION PROJECT : MAPPING AND TRANSFORMATION DESIGN OF YANSHE

**80**
建筑设计研究（一）南京老城南传统院落改建研究
DESIGN STUDIO 1  RENOVATION OF THE NORTH SIDE OF SHENGZHOU ROAD AND BOTH SIDES OF PINGSHI LAND

**88**
建筑设计研究（二）水乡聚落设计研究·呼吸作用
DESIGN STUDIO 2  WATER-VILLAGE OF RESPIRATION

**96**
建筑设计研究（二）城市空间设计·层与界
DESIGN STUDIO 2  FROM VOLUMN TO SPACE·MUTILLAYER & INTERFACE

**110**
建筑设计研究（二）建构研究
DESIGN STUDIO 2  CONSTRUCTIONAL DESIGN

**119—129**  建筑设计课程 ARCHITECTURE DESIGN COURSES

**131—133**  建筑理论课程 ARCHITECTURAL THEORY COURSES

**135—137**  城市理论课程 URBAN THEORY COURSES

**139—141**  历史理论课程 HISTORY THEORY COURSES

**143—145**  建筑技术课程 ARCHITECTURAL TECHNOLOGY

**147—153**  其他 MISCELLANEA

教学论文
ARTICLES ON EDUCATION

# 关于"居住建筑设计原理"
## TEACHING ON RESIDENTIAL BUILDING DESIGN PRINCIPLES

刘铨　冷天

根据最新的《高等学校建筑学本科指导性专业规范（2013年版）》（以下简称《专业规范》）的要求，建筑学本科阶段的设计理论学习包括了公共建筑、居住建筑、城市设计、室内设计原理四门课。而公共建筑与居住建筑作为建筑类型的两大门类，其设计原理更是历来作为建筑学本科阶段必修的课程。其中，居住建筑是与人们日常生活最为密切的建筑类型，从原理上剖析居住建筑的设计，可以更好地使学生把握设计的基本社会价值评判标准。因此，居住建筑设计原理在本科阶段的设计理论教学中占有非常重要的地位。

随着社会的发展，特别是房地产行业的兴起，居住建筑越来越多样化，与市场的结合也越来越紧密，因此居住建筑设计原理的教学理念、内容、方法也必须随之改变。笔者将结合南京大学建筑与城市规划学院开设的"居住建筑设计与居住区规划原理"课程，对此进行分析论述。

### 1. 教学理念：从设计到原理

目前适用于居住建筑设计原理的教材，都主要是按照类型来划分教学内容的，如住宅建筑、公寓建筑、居住综合体建筑，或是低层、多层、中高层与高层住宅等。在《专业规范》对课程知识点的安排中也体现出这一特点。这使这门课更像是建筑设计的辅助讲解课程，而不是一门设计理论课。而且，教材所列举的类型不仅不能涵盖面广量大、丰富多变的居住建筑，更重要的是，不同居住建筑类型之间的相似性与差异性所隐含的基本建筑原则，才是原理课应当教授的重点。因此，我们打破了传统教材的分类法，以设计的思维来划分教学内容，通过从单个功能空间、套型到单元、单体再到表皮与技术的逐步讲解，让学生从空间组织的角度体会居住建筑类型的相似性与差异性。

同时，我们将政治、经济、文化、生活（环境）作为贯穿的线索来讲解空间类型生成的影响因素。学生知其然更知其所以然，才能真正提高对设计问题的提取与分析能力，从而达到设计理论课的教学目的。因此，除了传统的知识点，我们在课程特别加入了以下几个方面的内容：（1）居住建筑历史发展的介绍；（2）相关设计理论发展的介绍；（3）社会经济文化因素，特别从住房分配到房地产市场化的转变对居住建筑设计的影响；（4）相关法规、规范对居住区规划与居住建筑设计的影响。例如，教学中我们以不同地域的民居建筑为载体，分析了建筑材料形式与地理气候的紧密关系，从政治、经济角度分析了新中国成立以来集合住宅单元平面的演变的阶段与过程，在讲解集合住宅单元类型时，着重讲述了法规、规范对类型塑造的重要影响，这些都帮助学生更好地理解了经济性、舒适性与强制性的功能要求在建筑设计中的相互制约关系。

### 2. 教学内容：从单体到规划

建筑单体的设计中，城市环境与系统成为越来越重要的考虑因素，居住建筑设计也不例外。在设计实践中，居住建筑单体设计大多也都与居住小区的修建性详细规划结合在一起。许多高校的本科建筑设计课程中也常常将住宅小区规划与住宅单体设计结合起来作为一个高年级的设计题目。这就要求我们在居住建筑原理课上也必须加入相应的内容。

同时，本科高年级学生要开始更多地接触城市空间，必然涉及城市规划与城市设计的知识，居住区规划的内容也可以成为这些知识与之前单纯的建筑设计相衔接的桥梁。住宅小区作为一个较为简单的城市空间类型，其中的用地与建筑布局、交通系统组织、绿地景观系统组织与设计、竖向与管线设计等知识，对于刚刚接触城市层面问题的学生会更容易掌握一些，因此和居住建筑单体知识也能够更好衔接，放在同一门课中也就顺理成章。

但在《专业规范》中，与居住区规划相关的知识点并不包括在居住建筑设计原理课程所要求的24学时教学内容之中，而是部分包含在城市设计原理（城市设计的典型种类之一，约1学时）和城乡规划原理（城市修建性详细规划，8学时）之中的。如果按照《专业规范》的学时安排，按每周3学时排课，上课半学期，每周的课程会过于冗长，知识点过多，而按照每周2课时排课，12周的课程又与通常一学期（16—18周）的时间安排不协调。因此，将居住区规划原理的内容纳入居住建筑设计原理课中，就可以按一整个学期、每周2课时的进度排课，形成从单个功能空间、套型、单元、单体到组团、小区这样一个尺度较为全面的居住建筑设计教学框架。

### 3. 教学方法：从授课到实践

原理课教学形式主要是课堂讲授。但这样的教学对于设计专业的学生来说较为被动，因为不进行实际操作就很难将理论内容转化为实际的设计经验。虽然许多建筑院系有住区主题的课程设计，但也很难在时间上与原理教学的课程安排相匹配。因此，单纯的授课教学，效果会大打折扣。正是基于这样的考虑，我们除了安排正常的授课，也设置了课外调研、课堂练习、讲座等形式的教学环节，形成知识反馈，以加深学生对相关知识点的理解。

课外调研。随着房地产市场的发展，住宅小区规划设计产品已经十分丰富多样，这些案例是学生学习的最好素材，也与他们的生活最为贴近。在课程开始前，课程会要求每个学生利用假期选择一处与自己居住地相近的住宅小区，完成一个调研表格，这样就建立了对居住建筑的初步认知。同时，所有学生的调研成果汇集为一个类型丰富的基础资料库，这样，学生在教学的不同阶段都可以利用其进行分析与归纳总结，如功能空间尺度分析、套型功能组织类型分析、小区总平面交通组织分析、绿地景观系统分析等，借助现实案例的分析对课堂教学的知识点进行反馈（图1）。

课堂练习。课堂练习的目的是训练学生将课堂理论教学的知识点、对调研案例的分析转化为设计思维的能力。因此练习都是按照授课顺序和住宅户型要素构成分步骤进行的：厨房设计练习，理解功能空间使用的规律、尺度、经济性；住宅套型与核心筒设计，理解舒适性、经济性与安全性要求的相互关系；单体立面设计练习，着重解决开窗面积比例、遮阳、空调机位等，理解功能与美观的关系；住宅小区总平面设计，理解规划的相关知识与要求。这些练习尽量安排在课堂规定时间完成，不挤占课外时间，同时要求徒手绘图，因此兼具快速设计训练的作用。

课堂讲座。讲座的目的是拓宽学生的思路，因此会邀请住宅设计、房地产开发与规划管理单位经验丰富的人员，从各自的职业角度来讲住宅的设计、产品研发与规划控制管理。通过讲座的互动，让本科的学生也有机会从不同的立场思考设计问题。

总体来看，我们不希望将"居住建筑设计原理"仅作为一门教授具体建筑类型设计知识的课程，而是将"居住建筑"这一与学生日常生活经验最为密切的建筑类型作为载体，让他们更加全面地认识"设计"需要考虑的基本要求，将设计与生活、与社会、与城市结合起来。同时，作为一般在三、四年级开设的课程，这一阶段的学生亦应当从主要关注单体的设计转向考虑更大尺度、关系更为复杂的群体建筑、城市空间的相关知识，因此这门课所讲授的城市及住宅与住宅区也应当成为一个很好的进行知识过渡的载体。

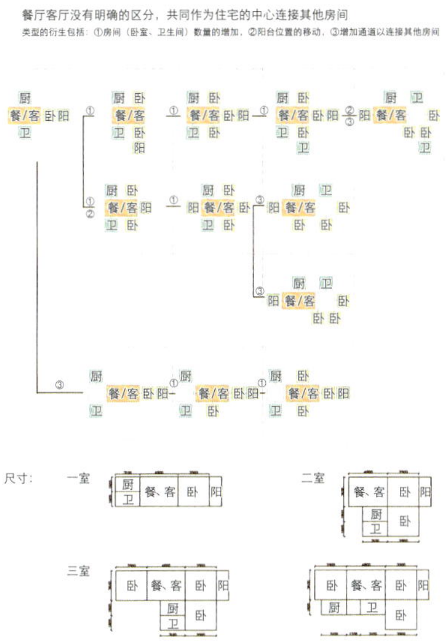

Figure 1 The typological analysis of living units by student

According to the latest requirements of Guiding College Norms for Undergraduate Courses o f Architecture (2013 edition) (hereinafter referred to as the Norms), the theoretical courses of architectural design at the undergraduate level includes the Principles of Public Buildings, the Principles of Residential Buildings, the Principles of Urban Design and the Principles of Interior Design. The Public Building and Residential Building is two major building types, so their design principles have always been viewed as the basic knowledge in the architectural design education of undergraduate level in China. Compared with the public building, Residential Building is more related with people's daily lives, and the theoretically analyzing of residential building can make undergraduate students better understand the design rules not only from the form and activity, but also from economy and society point of view. Therefore, the design principles of residential building take up a very important position in the teaching of the undergraduate level.

With the development of society, especially the rise of the real estate industry, residential buildings are more and more diversified and more tightly combined with the market. Therefore, the teaching concepts, contents and methods of residential building design principles will also be changed.

**1. Teaching idea: From design to theory**

The contents of most current teaching materials for the course mainly discuss the residential buildings by different types, such as housing, apartment, residential complex buildings, or lower-rise, multi-rise, and high-rise buildings, which is also reflected in the Norms. It makes the course more like an auxiliary explanation course for building design, but not a design theory course. Besides, the types illustrated in the course cannot cover the abundant residential buildings of great varieties, more importantly, the basic building principles hidden in the similarities and differences among different residential building types are the focus of teaching in the theory course. Therefore, we break the classification method of traditional teaching materials, divide the teaching content by the design thought and explain step by step from the single functional space and dwelling size to units, monomers, cuticles and technologies, to make students experience the similarities and differences of residential building types from the perspective of spatial organization.

Meanwhile, we take politics, economy, culture, life (environment) as the penetrating clue to explain the influential factors of spatial type generation. Only by knowing what and why it is can students really improve their extraction and analytical ability of design issues, so as to achieve the teaching aim of the design theory course. Therefore, in addition to traditional knowledge points, we specially add the following contents to the course: (1) the brief history of residential building;

(2) the related design theory; (3) the social, economic and cultural factors, especially the influence of the transformation from housing distribution to real estate market; (4) the influence of the relevant codes and regulations on residential building design and residential district planning. For example, we analyze the close relationship between the building material form and the geographical climate from different regions of residential buildings as the carrier and analyze the evolution stage and process of amalgamated dwelling unit planes since the founding of new China from political and economic angles. While explaining the types of amalgamated dwelling units, we focus on teaching the significant influence of laws, regulations and standards on type molding. These all help students better understand the mutual restrictive relations of economical, comfortable and mandatory functional requirements in building design.

## 2. Teaching contents: From building to planning

The urban environment becomes more and more important factors in the design of single buildings, so as the residential building design. In design practice, the designs of single residential buildings are mostly combined with the site plan of residential quarters. The undergraduate architectural design courses in many colleges also often combine residential quarter planning with single residence design as a senior design subject, which requires that we also must add corresponding contents to the course of the residential building design principle.

Meanwhile, the senior students of undergraduate start to contact with urban spaces, which inevitably involve the knowledge of urban planning and urban design. The residential quarter is a simple urban space type, so students can be easier to understand the basic urban planning knowledge, such as land use and building composition, the organization of transportation system, landscape design, and vertical and pipeline design. The residential quarter planning also can better link the previous knowledge of single building design, so it is reasonable to put it into the course.

However, in the Norms, the knowledge points related to residential planning are not included in the 24 class hours of the course, but are partially included in the urban design principle (about 1 class hour) and urban-rural planning principle (8 class hours). If it is done in accordance with the class hour arrangement of the Norms, with 3 class hours per week in a half semester, the weekly courses will be long-winded and the knowledge points imparted will be too numerous, while with the arrangement of 2 class hours a week, the 12-week course will be very discordant with the time arrangement of a semester (16-18 weeks). Therefore, if the contents of residential planning principle are brought into the course, two class hours a week can be arranged in a whole semester and form a more comprehensive residential building design teaching framework ranging from the single functional room, living unit, dwelling unit, single building to residential clusters and quarters.

## 3. Teaching method: From teaching to practice

The instructional method of the theory course is mainly classroom instruction. But it is passive for design major students, for it is very difficult to convert the theoretical contents to practical design experience without practical operation. Although many architectural departments have residential themed curriculum design, it is very hard to match with the course arrangement of time. Based on such consideration, except for the normal teaching, we also set up extracurricular surveys, class exercises, lectures, and other forms of teaching methods to form knowledge feedback and deepen students' understanding of related knowledge points.

### Extracurricular survey

With the development of the real estate market, the housing products have been very abundant and diversified. These cases are the best materials for students' learning and are closest to their lives. Before the start of the course, every student needs to choose a residential quarter similar to their place of residence to complete a survey during their holiday to set up their preliminary understanding of residential buildings and quarters. Meanwhile, all students' survey results are brought together as a basic data bank with rich types, and the students can utilize it to make analysis and summarization at the different stages of teaching, such as the analysis of functional space dimension, functional organization pattern of living units, traffic organization of the residential quarter, and green space landscape system, which can respond to the knowledge points from classroom instruction.

### Practice sessions in or after class

The objective of these practice sessions is to train students' ability to transform the knowledge points of classroom instruction and analysis of survey case to design thinking, so practice sessions are conducted by steps according to the order of teaching and the of residential space components: kitchen design aims to understand the usage, scale and economic efficiency of functional space; living unit and core tube design aims to understand the relationship between comfort, economic efficiency and safety requirements; single building façade design focuses on the relationship between functions and aesthetics through solving window area proportion, sun shading, air-conditioner space; the residential quarter site planning aims to understand the related knowledge and requirements of planning. These practices should be arranged to be completed within class time as much as possible and not occupy extracurricular time.

Lectures

The purpose of lecturing is to broaden students' thinking, so people with rich experience in different positions will be invited to the course , such as designer, real estate developer and official of planning management, to give the students different perspective of design. Tthrough the lecture interaction, the undergraduate students can rethink the design value from another standpoints.

Overall, we hope the course of "Residential Building Design Principle" is not just a course for teaching the design knowledge of specific building type, but to take "residential building" as a carrier closest to students' daily life experience, so as to make students more comprehensively recognize the basic requirements that should be considered in "design" and combine design with life, society and city. Meanwhile, as a course set in third and fourth years, the students at this level should also turn to consider the knowledge of large scaled and complex urban space from mainly focusing on the design of single buildings. Therefore, the city, residence and residential quarter taught in the course should also become a very good carrier of knowledge upgrade.

# 以城市物质形态为基础的本科城市设计理论课程教学研究
A PEDAGOGICAL STUDY ON THE COURSE OF URBAN DESIGN THEORIES FOR UNDERGRADUATE STUDENTS, BASED ON URBAN PHYSICAL FORMS

胡友培　丁沃沃

在本科高年级阶段（四年级上），安排"城市设计及其理论"的课程教学，是南京大学建筑学专业本科生培养计划的重要一环。该理论课程设定为16讲（32学时），教学目的是让高年级本科生初步了解城市设计的基本内容，掌握有关概念、理论与方法。该课程与一个城市设计大作业相平行，训练学生融汇理论与实践的能力。从更长远的培养计划出发，将城市设计课程设置在高年级阶段，其意图是拓展专业视野，培养城市建筑（urban architecture）的基本观念，为后续研究生阶段的学习做好准备与铺垫。本文将对该课程组织架构进行简要介绍，重点讨论其中的基础理论部分，并围绕相关问题展开讨论思考。

### 1. 城市设计理论课程构架

以何种架构选择并组织16讲的教学内容，将基础落在哪个方面，是该课程面对的第一个问题。有关城市设计的理论性著述，可以列出一个长长的书单。如较早的卡米诺·西特的《遵循艺术原则的城市设计》，20世纪中后期的《美国大城市的死与生》《城市意象》《好的城市形态》，以及当代的 Urban Design since 1945 : A Global Perspective、Shaping the City: Studies in History, Theory and Urban Design 等。这些著作构成了西方城市设计理论的基石。在中文书籍里，王建国老师的《城市设计》对我国城市设计影响广泛。此外还有卢济威老师的《城市设计创作：研究与实践》、段进老师的《空间研究》系列等。这些中文著作积极引介西方理论的同时也致力于发展我国的城市设计理论话语。上述书籍是我们城市设计理论课程重要的文献资源。另一方面，每本著作都有其特定的写作时空背景，从中挑选一本书籍直接作为当下我国城市设计理论课程的教材，具有一定的难度。

在综合参考城市设计相关理论书籍基础上，结合教学对象与教学目标，我们将课程内容划分为几大方面：（1）课程概述；（2）城市物质形态理论；（3）城市设计的技术方法；（4）城市设计理论延伸。概述部分讲解城市设计的基本概念以及简介整个课程；形态部分讲授城市形态的基本内容与方面；方法部分主要包含设计方法、分析手段与技术语言；延伸部分在理论性上有较大提高，主要涉及城市设计理论的历史发展、城市设计的物理环境维度、城市设计的人文环境维度。显而易见，这是一个由综述、城市形态、技术方法与理论提升组成的渐进式的课程框架。

在上述几个方面中，将基础落在理论提升上显然不太适宜，因为这部分内容理论性、专题性较强，并不适合作为本科教学的基础内容。形态理论与技术方法，一个关于"是什么"，另一个关于"如何做"，其中具有较多的初等内容。在二者中选择任一作为课程的基础，都具有合理性。选择的依据取决于课程的基本导向。在导向实践的教学框架内，适宜以设计方法作为基础。考察我们的本科课程体系，由于有一个作为实践的城市设计练习与理论课平行，因而课程更偏向于理论导向，侧重于传授知识、概念与思维方法。由此，将课程的基础放在城市物质形态理论上。

### 2. 基础课程的内容构成

有关城市物质形态的理论应以何种理论形式呈现，以及由哪些内容构成，是该课程需要思考的第二个问题。凯文·林奇（Kevin Lynch）曾提出过三种类型的城市理论形态，分别为描述性、规范性与批判性理论。描述性的理论解决对象是什么的问题；规范性的理论导向对象应该如何的问题；而批判性理论则主要立足于揭示问题，引发思考。从理论的难度上，上述三种形态呈现出递进的关系。其中描述性理论侧重于客观性，规范性理论具有一定的价值判断，而批判性理论是思辨。林奇的这个分类法，为我们的问题提供了思路：面对本科生的理论课程是基础性质的。因而有关城市物质形态的理论采取描述的形式较为适宜。规范性的理论形式，涉及预设的价值判断；而批判性的理论则破坏性大于建设性。两种形式对于本科生而言，都存在过于复杂的弊端。而描述性理论较为直白，以客观陈述、讲解城市物质形态为主要内容，使得学生了解并理解城市设计的主要工作对象。这个理论解决的问题是：我们如何开口言说城市形态，或城市形态是什么？都有哪些组成和结构？如何进行分类，以及随之而来的概念和知识。

在确定以城市形态描述性理论作为基础后，其具体内容在构成上也需要进行一番取舍。课程大致从四个方面对城市形态展开描述——结构、肌理、路网与空间。其中包含着一个由大到小、由粗到细的递进尺度关系。试图帮助学生建构一个较为全面的、具有一定系统性的知识体系。

#### 2.1 城市形态的结构与机制

课程中所谈论的城市，主要针对城市（city），而非规划尺度上的都市圈（metropolitan）。讲座内容并不过多地涉及城市的抽象结构（如芝加哥学派提出的经典城市结构——同心圆、多中心、扇形等），而被限制在较为具体的结构形态上。即，"结构"仍然是基于物质形态的。著名的城市学家 Spiro Kostof 的有关著作是这部分课程的主要理论来源。Kostof 援引林奇的分类，将城市的结构形态划分为宗教城市、自然有机城市、格网城市。进而提出相应的城市形态的生成机制，如：宗教城市反映了绝对权力意志对城市的塑造；自然有机城市是自下而上的；而网格城市在工业革命后大量涌现则是出于效率和经济的原因。这个分类并不是唯一的，甚而并不全面。但一方面该分类胜在精炼，更重要的是指出了形态与机制的紧密联系。这一点对于学生理解城市形态尤为重要：不仅知其然，更知其所以然。

#### 2.2 城市路网与交通

在城市设计理论中，讲授有关交通、路网的知识，我们将重点仍然放在形态方面——城市的路网模式及其形态。在初步了解交通工程的道路分类与术语基础上，主要的内容是让学生建立一种基于形态的城市路网观念，学会从形态、构型的角度去理解、描述路网。在理论资源上，主要援引希列尔 Bill Hilliuer 的空间句法和 Stephen Marshall 的街道模式理论。课程并不深入空间句法的技术层面，而主要将其作为一种理解城市路网复杂结构的理论工具，帮助学生从构型（configuration）的层面去理解城市形态中的一个重要维度——网络维度，学习描述网络的一些基本词汇，如节点、边、深度、构型等。马歇尔的著作在对路网的描述方法论上，与空间句法有共通之处。其独特之处在于展示了各种经典路网模式及其量化描述，使得学生在掌握基本"形"的同时，了解到"量"的存在维度。

#### 2.3 城市肌理与方法

城市肌理是中观尺度下最为普遍的城市形态。与之属于同一层次的还有城市景观、基础设施、节点性建筑等。它是城市设计的普遍对象，建筑设计的普遍背景。课程主要讲授城市肌理的基本内涵，以及相应的描述方法。城市形态学（urban morphology）的相关理论是这部分课程的理论资源。著名城市形态学家M.R.Conzen提出了城市形态由平面单元、建筑、地块3个层级组成；在此基础上，发展了城市肌理形态的3分描述框架：街区、地块、建筑。该描述框架易于理解和操作，是有关城市肌理的基础理论工具。除了空间尺度外，课程还引介了意大利形态学派的类型—形态学方法，从类型演化的角度解释城市肌理的变迁，帮助学生拓展在时间维度上对城市形态的认知。进一步，课程将这套描述工具应用于具体案例（如巴塞罗那的方格网街区肌理、曼哈顿的摩天楼街区肌理、我国城市的大街区肌理、我国历史城市肌理等），让学生在应用中对其加以掌握。

城市路网与城市肌理，二者在现实中是紧密结合的。在拆分以便讲授的同时，需要培养学生形成一种综合的意识。在肌理的教学中，课程会不断地返回路网的相关知识，以对肌理的形态做出分析和描述。

2.4 城市空间与形态

空间是建筑学专业学生最容易理解的词汇。然而，在城市层面，学生有关城市空间的概念相对薄弱，需要培养其形成城市空间的概念以及掌握城市空间的基本词汇。在课程中主要讲授几种常见的城市空间类型——街道空间、广场空间、景观绿地空间、滨水空间等。有关西方城市空间的基础理论相对丰富。罗伯特•克里尔的理论帮助学生进入欧洲古典城市，库哈斯、文丘里等的理论则提供对当代西方城市的精彩解读。对于我国当代城市的空间概念与词汇，主要来源于城市规划相关术语，如公共空间体系、城市绿地系统以及较为宽泛的各种轴、廊、节点等。需要指出的是，由于城市规划的术语并不足以反映我国城市空间在城市设计层面的生动性与多样性，因而还需加强这部分的基础研究，以支持基础课程的教学。

3. 讨论

（1）城市设计的理论浩瀚如烟，不可能仅以一个框架覆盖之。我们的本科教学所采取的框架以本科生为对象，以传授基础知识为主要目的。复杂如批判性的城市理论并未列入其中。这些理论可以留待研究生阶段再行专题研讨。

（2）我们所采取的框架将基础放于城市物质形态的描述性理论方面。需要再次指出的是，有关城市设计理论的基础选择并非唯一。选择的依据取决于课程的基本导向。在导向实践的教学框架内，以设计方法作为基础，同样具有合理性。另一方面，该基础的理论形式也并非只有描述性一种。诸如规范性与批判性的理论形式，由于涉及价值基础与思辨，更适宜作为研究生教学的理论形式。我们认为描述性理论，在帮助学生形成较为坚实的知识基础同时，为今后自主的发展留有空间。

（3）城市物质形态理论的具体内容，具有多种构成与组织的可能。我们的课程仅是其中一种。还可以从地域角度、历史时间维度等方向进行组织。在教学中，课程除了帮助本科生较顺利地掌握具体知识外，还试图传递一个具有层级递进性质的知识体系，以便于掌握进一步的知识。

At the senior undergraduate stage (the first semester of the fourth grade), arranging the course teaching of Urban Design and Theories is an important link in the cultivation plan of students majoring in architecture at Nanjing University. The theory course sets 16 lessons (32 class hours), and the teaching objective is to make senior undergraduate students have a preliminary understanding of the basic contents of urban design and master relevant concepts, theories and methods. The course is parallel to a big assignment of urban design to train students' ability to combine theory with practice. From the perspective of longer cultivation planning, the intention of setting the urban design course at the senior grade stage is to extend professional horizons, cultivate the basic concepts of urban architecture, and make preparations to pave the way for learning at the follow-up graduate stage. The paper briefly introduces the organizational structure of the course and focuses on discussing the basic theoretical parts within and reflecting on related problems.

**1. Framework of the Course of Urban Design Theories**

By what framework the teaching contents of the 16 lessons are selected and organized and at which aspect the foundation is laid is the first problem facing the course. A very long book list can be listed for theoretical works about urban design, such as the earlier Camillo Sitte's *Urban Design Following Artistic Principles, The Life and Death of American Metropolises, City Image, and Good Urban Forms* in the middle and later periods of the last century, as well as *Urban Design Since 1945 : A Global Perspective, Shaping the City: Studies in History, and Theory and Urban Design*. These works constituted the theoretical cornerstone of Western urban design. In Chinese books, Professor Wang Jianguo's *Urban Design* has a wide influence on urban design in China. Additionally, there is Professor Lu Jiwei's *Urban Design Creation*: Research and Practice, Professor Duan Jin's *Spatial Research*, and so on. These Chinese works actively introduced Western theories and are also devoted to developing urban design theoretical discourse in China. The books above are the important literature resources of our course in Urban Design Theories. On the other hand, each book has its specific writing background, so it is difficult to directly select a book as teaching material for an urban design course in China. On the basis of comprehensively referring to the related theoretical books of urban design and combining the teaching object and teaching goals, we divide the course contents into the following aspects: (1) course overview, (2) urban physical form theory, (3) urban design technical methods, (4) urban design theoretical extension. The overview part explains the basic concepts of urban design and briefly introduces the whole course, the form part teaches the basic contents and aspects of urban forms; the method party mainly includes design methods, analytical means and technical terms; the extension part is greatly improved theoretically, mainly involving the historical development of design heories, physical environment theories, physical environment

图1 城市形态结构课件内容摘选
Figure 1　Urban form structure courseware content selection

图2　路网与交通课件内容摘选
Figure2　Road network and traffic courseware content selection

dimension of urban design and the cultural environment dimensionality of urban design. Obviously, this is a progressive course framework consisting of overview, urban form, technical methods and theoretical improvement.

In the above-mentioned aspects, it is obviously improper to lay the foundation on theoretical improvement, because part of contents is of stronger theoretical property and topicality, and is not suitable as the basic contents for undergraduate teaching. Form theories and technical methods, one about "what is it" and the other about "how to do it", have many elementary contents. It is rational to select any one of the two as the foundation of the course. The basis of selection lies in the basic orientation of the course. Within the teaching framework of guiding practice, it is suitable to take design methods as the foundation. In our undergraduate course system, there is practical urban design practice in parallel with the theoretical course, so the course is more inclined to theoretical guidance and lays more emphasis on imparting knowledge, concepts and thinking methods. Therefore, the foundation of the course is laid on urban physical form theories.

**2. Content Construction of Basic Courses**

By what theoretical forms the urban physical form theories are presented and what contents they are made up of is the second problem that should be reflected in the course. Kevin Lynch once put forward three types of urban theoretical forms – descriptive, normative and critical theories. The descriptive theories solve the problem of what is the object; the normative theories guide the problem of how the object should be; the critical theories mainly focus on revealing problems and triggering thoughts. From the difficulty of theories, the above three forms are of progressive relations. The descriptive theories lay emphasis on objectivity, normative theories have a certain value judgment and critical theories are speculation. Lynch's classification method provides ideas for our problems: The theoretical course facing undergraduate students is fundamental. So the descriptive form is more suitable for the theories related to the urban physical form. The normative theoretical form involves presupposed value judgment, while critical theories have more of a destructive than constructive effect. The two forms both have overly complex drawbacks for undergraduate students. Descriptive theories are more straightforward and take objective statements and urban physical form explanation as their main contents, leading students to learn and understand the main work objects of urban design. The problems solved by the theories include how to talk about urban forms, what urban forms are, what composition and structure they have, how to make classification, along with the consequent concepts and knowledge.

After confirming the urban form descriptive theory as the foundation, its specific content components should also be selected. The course roughly describes urban forms from four aspects – structure, texture, road network and space, and includes a progressive dimension relationship from big to small and from coarse to fine, trying to help students build a more comprehensive and systematic knowledge system.

2.1 The Structure and Mechanism of Urban Forms

The cities discussed in the course mainly target cities, not metropolitan areas, in terms of their planning scale. The teaching contents do not excessively involve urban abstract structure (such as the classical urban structures put forward by Chicago School – concentric circles, multicenter, sector, etc.), but are limited to more specific structural forms, namely, "structure" is still based on physical forms. Famous urbanologist Spiro Kostof's related works are the theoretical source of this part of the course. Kostof cited Lynch's classification and divided urban structure forms into religious cities, natural organic cities and grid cities, and then raised the corresponding urban form generative mechanism, such as religious cities reflecting the molding of the power will to cities; natural organic cities are from bottom to top; the springing up of grid cities after the industrial revolution was for efficiency and economic reasons. The classification is not singular, and so is not complete, but it does succeed in refinement. More importantly, it pointed out the close connection of forms and mechanism, which is particularly important for students to understand urban forms: not only knowing how but also knowing why.

2.2 Urban Road Network and Traffic

In urban design theories, we will still focus on form – urban road network patterns and forms impart knowledge about traffic and road networks. In the preliminary understanding of the road classification and terminology basis of traffic engineering, the main content is to let students build up a kind of urban road network concept based on forms and learn to understand and describe road networks from the perspective of forms and architecture. In terms of theoretical resources, Bill Hilliuer's space syntax and Stephen Marshall's street pattern theory are mainly quoted. The technical level and space syntax does not penetrate throughout the course, but is a

图 3 城市肌理课件内容摘选
Figure 3  Urban texture courseware content selection

图 4 城市空间课件内容摘选
Figure 4  Urban space courseware content selection

theoretical tool to understand the complex structure of urban road networks. It helps students to understand network dimensions, an important dimension in urban forms from the configuration level. They learn to describe some basic terms of networks, like node, side, depth and configuration. Marshall's works have something in common with space syntax in the description methodology of road networks. Its uniqueness lies in showing various classical road network patterns and their quantizing description, so as to help students understand the existing dimension of "quantity" while mastering the basic "form".

2.3 Urban Texture and Methods

Urban texture is the most common urban form in the mesoscale. Things at the same level include urban landscape, infrastructure, node buildings, etc., and it is the common object of urban design and the common background of building design. The course mainly teaches the basic contents of urban texture and the corresponding descriptive methods. The related theories of urban morphology are the theoretical resources of this part of the course. Famous urban morphologist M.R.Conzen put forward the urban form consisting of plane units, buildings and land parcels; on this basis, three descriptive frameworks of urban texture forms are developed: Street block, land parcel and building. The descriptive framework is easy to understand and operate and is the basic theoretical tool related to urban texture. Besides spatial scale, the course also introduces the morphological method, the form of Italian form school, explains the transition of urban texture from the perspective of type evolution and helped students expand their cognition on urban forms in the time dimension. Furthermore, the course applies the set of descriptive tools to specific cases (such as Barcelona's grid square block texture, Manhattan's skyscraper block texture, the big street block texture of Chinese cities and Chinese historical city texture), to help students grasp it in application.

Urban road network and urban texture are closely combined in reality. In splitting them to facilitate teaching, it needs to cultivate students to form a kind of comprehensive awareness. In texture teaching, the course will constantly return to the related knowledge of road networks and make analysis and description on the forms of texture.

2.4 Urban Space and Forms

Space is a word that architecture students can understand most easily. However, at the urban level, students have relatively weak concepts of urban space and need to cultivate the formation of the concept of urban space and master the basic terms of urban space. A number of common urban space types are mainly taught in the course, including street space, square space, landscape greenbelt space and waterfront space. The basic theories about Western urban space are relatively richer. Robert Kerry's theories help students enter into European classical cities; the theories of Koolhaas and Venturi provide wonderful explanations of modern Western cities. The space concepts and words about Chinese contemporary cities are mainly from the related terms of urban planning, such as public space system, urban greenbelt system, and various axles, corridors and nodes. It must be pointed out that urban planning terms are insufficient to reflect the vitality and diversity of Chinese urban spaces at the urban design level, so the basic research on this part still needs to be strengthened to support the teaching of basic courses.

3. Discussion

(1) Urban design theories are extremely abundant, and it is impossible to cover them with a framework. The framework adopted in our undergraduate teaching targets undergraduate students, mainly for the purpose of imparting basic knowledge. The complex critical urban theories are not listed within. These theories can be reserved for graduate students to study and discuss.

(2) The framework adopted by us puts the foundation in the descriptive theories of urban physical forms. It must be pointed out again that the selection concerning the urban design theoretical foundation is not singular. The basis of selection depends on the basic guide of the course. In the teaching framework of guiding practice, it is also reasonable to take design methods as the foundation. On the other hand, the theoretical form of the foundation also does not just contain the descriptive form. The normative and critical theoretical forms, involving value basis and speculation, are more suitable as the theoretical form for graduate teaching. We think descriptive theories leave room for future independent development while helping students form a more solid knowledge foundation.

(3) The concrete contents of urban physical form theories have the possibility of multiple constitutions and organizations. Our course is just one kind. It can be organized from regional perspective, historical and time dimension and other directions. In teaching, the course not only helps undergraduate students smoothly master specific knowledge but also tries to convey a knowledge system with a hierarchical progressive nature, so as to further master knowledge.

# "建筑环境学"教学学生科研能力培养的探索
EXPLORATION INTO THE DEVELOPMENT OF STUDENTS' RESEARCH CAPABILITIES IN THE 'BUILDING ENVIRONMENT' COURSE

郜 志

南京大学建筑与城市规划学院于2013年春季开始开设了"建筑环境学"的硕士研究生课程。该课程涵盖传统意义的"建筑环境学"并涉及城市尺度的"城市物理环境学"，主要关注建筑、城市与外部环境、与室内环境及与人之间的关系，内容覆盖热学、流体力学、物理学、生物学、气象学、心理学、生理学、卫生学等学科知识。目前主要授课对象为建筑技术方向（必修）与建筑设计、城市设计方向（选修）的一年级硕士研究生。由于该课程涉及室内空气品质与污染的内容，2014年春季也有生命科学学院等其他学院的研究生选修此课。

### 1. 课程安排

为提高教学的有效性，尝试将课程安排分为"建筑环境学"基础理论知识的授课及建筑环境专题研究，以调动学生学习的积极性。"建筑环境学"是一门18学时（1学分）的课程，按照基础理论授课、专题选题、PPT讲解与答辩、论文提交等几个步骤进行课程安排。需要指出的是第9个周末需提交论文初稿，经意见反馈后于第18周提交终稿。基础理论的教学与开放式研究课题的具体安排如下：

#### 1.1 基础理论教学

该课程的主要任务是使学生掌握建筑环境的基本概念，学习建筑与城市热湿环境、风环境和空气质量环境的基础知识。通过课程教学，使学生熟悉建筑环境与城市微气候等理论，并了解人体对热湿环境的反应，掌握建筑环境学的实际应用和最新研究进展，并为建筑能源和环境系统的测量与模拟打下坚实的基础。基础理论的授课包括如下内容：

建筑环境、城市物理环境、太阳辐射；
建筑热湿环境、冷负荷与热负荷计算原理与方法、城市热湿环境；
人体对稳态和动态热湿环境的反应及数学模型、室内室外热舒适性、热环境与劳动效率；
建筑自然通风与机械通风、城市风环境、室内室外风环境的评价指标；
室内空气品质、城市大气质量、空气质量对人的影响及评价方法、污染控制方法；
建筑与城市声光环境基础；
建筑与城市环境测试技术。

#### 1.2 开放式研究专题

"建筑环境学"开设在研究生一年级的第二学期，处于硕士学位论文开题之前非常重要的学习和过渡阶段。课程教学目的不仅仅是给学生传授建筑环境的基础理论知识，更希望通过这门课程，培养学生的科研能力，为今后学位论文的开展打下坚实的基础。为此，课程结合课程理论与学科研究热点，专门设计了一些研究专题供学生们选择和尝试，进行专门性研究（表1）。研究专题是开放性质的，学生还可以根据自己的兴趣，或未来学位论文方向自行确立专题题目。

### 2. 教学实践讨论

"建筑环境学"是建筑环境与设备专业一门重要的专业基础课。南京大学建筑技术科学专业成立的时间不长，大部分建筑技术专业学生之前的专业背景为建筑设计方向。这为课程的教学带来巨大的挑战，但同样也是教学研究的一次重大机遇。填鸭式的教学事倍功半，学生的主观能动性需要激发。故本课程尝试将基础理论知识与研究专题相结合的教学方式，希望调动学生的学习积极性，并培养其科学研究能力。

#### 2.1 教材精读

虽然"建筑环境学"对培养学生的科研能力进行了积极的探索，但需要引起注意是基础理论的教学依然是本课程的核心目的，不能舍本逐末。而学生对基础理论的掌握，仍然需要以对教科书的精读为主要手段。为调动学生的精读积极性与有效性，课堂上对学生进行以下类似问题的提问，比如：

计算夏季的冷负荷能否采用日平均温差的稳态算法？
环境空气温度、平均辐射温度和操作温度的关系是什么？
空气龄、残留时间和驻留时间的关系是什么？
……

"建筑环境学"的教学一直贯穿加深学生对建筑与城市的"舒适性、健康、节能、对环境影响及经济性能"等方面的综合认识，并强调室内室外关系的考量，比如舒适性和通风等室内外评价指标、室内外污染物传播规律等。

#### 2.2 科技文献检索与阅读

科研能力的培养首先依然应建立在对教材的掌握之上，其次需增强文献的阅读能力，并采用科学合理的流程。比如对于某项绿色建筑技术来讲，要搞清它的来龙去脉，需要结合教材和文献检索，如该项技术国际国内应用情况，不同气候区、不同地区的使用情况，最早使用情况，最成功案例、失败案例及主要原因，对环境的影响，技术评价的现有计算方法及已有软件等。

文献的检索与阅读是科研工作必备的能力之一，以是否有"扬灰层"（"扬尘层"）问题为例来说明如何提高学生文献阅读与分析能力。从2003年起，网络就有

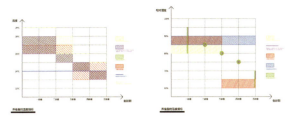

图1 养殖蚕的温湿度环境分析（蚕农场建筑设计研究分析，学生：孙昕、曹政）
Figure 1 Analysis on the Temperature and Humidity for Silkworm Breeding (Analysis and Study on Architectural Design of Silkworm Farms, by Sun Xin and Cao Zheng)

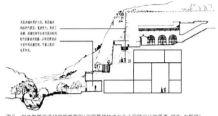

图2 相关聚落改进研究粗略案例（双层幕墙技术与乡土实践设计的思考，学生 力振球）
Figure 2 Case Study on Relevant Settlement Improvement (Thinking about Double-Skin Facade Technology and Local Practical Design, by Li Zhenqiu)

"售楼小姐真情告白"，其第6条"别以为高层中的九到十一楼不错，那你大错了，这些楼层正好是扬灰层，脏空气到这个高度就会停顿，我们是不会告诉你们的"这一说法，会影响到部分人群对不同楼层的购买意向，但是否具有科学道理？在文献检索过程中，有同学检索到摘要为"应用概率统计理论，对目前人们普遍关注的扬灰层问题构建模型，给出了在高层建筑附近，灰尘浓度随着高度变化的分布规律，并由此得出扬灰层存在的结论"的科技文章，信以为真并全盘接受。但通过课堂上对其深入的分析、讲解及讨论，学生最终意识到其分析手段存在很大问题，其结论并不可靠。而通过进一步的检索，找到报刊文章"本报记者实地测量发现，'9—11层是扬灰层'没根据，PM10、PM2.5浓度与楼层无关，与楼房所在区域的'微气候'有关"（人民日报，2013年7月2日第4版）。这篇文章虽不是严格意义的科技论文，但其采用的分析步骤合理，结论更为可靠。通过这个例子，学生意识到并不是所有的科技文献都是可靠的，需要从各个方面判断其准确及可靠性，并加以甄别与筛选，才能有助于今后科研工作的开展。

### 2.3 研究课题的逐层深入

学生科研能力的提高是一个相对漫长的过程，并不是一门"建筑环境学"所能解决的。但该课程中研究课题的设立和运作对学生科研创新能力的培养依然有其较为积极的作用。比如从选题开始，到期中的论文题目，再到最终的期末论文确立这一过程，除了少数学生从始至终思维分不变，大部分学生都有一个课题从宽泛到具体、从概念化到具体可操作化的一个过程。还有部分同学由于各种原因，抛弃原有选题（表2）。这些都是科研活动中常见的问题，学生尽早接触到这些，会为其日后的科研活动下良好的基础。

### 2.4 教学成果展示

经过一学期的"建筑环境学"学习和研究专题的开展，除了个别学生表示依然找不到科研的感觉，大部分学生均能结合相关专题熟练进行中英文科技文献检索，并将建筑环境学的理论知识结合到专题研究和建筑设计中去。图1展示了学生在文献调研后对具体建筑环境的分析结果，图2展示了学生考虑到当地环境因素的建筑环境设计。虽然由于课程时间的限制，学生专题的研究还处于较为粗浅的阶段，目前看来，仍取得了一定的成效。而如何跟踪学生是否真正将课程中所学内容及研究手段应用到之后的科研活动中，是未来需要考虑的教学课题之一。

表1 "建筑环境学"开放式研究专题列表

| 建筑环境与节能技术 | 风环境及空气质量 | 城市微气候 | 人体对热湿环境的反应 |
|---|---|---|---|
| 太阳能一体化技术 | 通风效率的指标：从室内到室外 | 城市微气候模拟软件 | 室内外热舒适性指标评价 |
| 建筑环境与节能技术 | 风环境及空气质量 | 城市微气候 | 人体对热湿环境的反应 |
| 太阳能集热墙太阳能烟囱 | 室内外空气质量的国家和行业标准比较 | CFD用于室外环境计算时的计算区域 | 热环境与劳动效率 |
| 双层皮幕墙 | 室内外空气污染物（悬浮颗粒，臭氧）的传输关系 | CFD模拟技术的准确性分析 | — |
| Low-E 玻璃 | 生物气溶胶特性研究 | 植物与树冠、车辆的模拟方法 | — |
| 光伏发电技术 | 空气净化器杀菌技术 | 多孔介质方法模拟城市微气候 | — |
| 垂直绿化与屋顶绿化技术 | 空气净化器效率 | | |
| 雨水回收系统 | 紫外灯杀菌技术 | | |
| 奥运场馆、世博园等临时建筑的环境评价 | 绿色植物净化空气效果 | | |

表2 "建筑环境学"研究课题的发展进程

| 选题 | 期中论文题目 | 期末论文题目 |
|---|---|---|
| 太阳能与建筑一体化技术 | 特伦布墙 | 传统特伦布墙合理尺寸与实例的构造方式及应用 |
| 太阳能与建筑一体化技术 | 双层玻璃幕墙 | 关于双层幕墙技术的研究进展 |
| "扬尘层"分析研究 | 双层皮幕墙 | 双层幕墙技术与乡土实践设计的思考 |
| 屋顶花园文献综述 | 屋顶绿化在改善室内以及城市环境的作用 | 夏热冬冷地区屋顶绿化问题探讨 |
| 垂直绿化与立面设计 | 垂直绿化的生态作用 | 垂直绿化的虫害防治 |
| 植物的数学模拟 | 城市微气候下树木的热环境模拟研究综述 | 利用计算机模拟技术评价植物对室外微气候影响的具体方法 |
| 建筑通风 | 室内通风的评价标准以及如何把室内通风引到室外 | 气流组织评价标准以及将室内评价指标应用到室外的分析研究 |
| 室内污染物：臭氧 | 臭氧污染特征与机理、浓度监测与评估方法及污染消除方法 | 地面大气臭氧浓度数据分析及臭氧浓度控制方法 |
| 室内空气微生物污染及其防治 | 紫外灯消毒效率及其产生的危害 | 紫外灯效率及其产生的危害 |
| 植物模拟与室内空气污染 | 植物在净化室内甲醛中的应用 | 绿色植物净化室内甲醛研究进展 |
| 室内空气污染及对策 | 室内空气净化器 | 负离子空气净化器 |
| 蚕农场建筑设计研究 | 蚕农场建筑设计研究分析 | 蚕农场建筑设计研究分析 |

The School of Architecture and Urban Planning, Nanjing University has delivered a graduate course on "Built Environment" since Spring 2013. Involving the traditional "built environment" and the "urban environment physics" in terms of urban scale, this course focuses on the relationship between buildings/cities and the outdoor environment/indoor environment/human occupants, and covers thermal science, fluid mechanics, physics, biology, meteorology, psychology, physiology, hygiene, etc. Currently, the students are 1st-year graduate students majoring in building technology (required), architectural design and urban design (elective). The graduate students from other schools such as the School of Life Science also took this course starting from Spring 2014, as it involves indoor air quality and pollution.

**1. Curriculum**

To improve the effectiveness of the teaching and motivate the students' learning initiative, we attempt to divide the course into one concerning the fundamental theoretical knowledge about built environment and the research project of built environment. "Built Environment" is a course containing 18 class hours, with the procedures of fundamental theoretical knowledge instruction, specialized study, PPT presentation and defense, and final report submission. It should be noted that the first draft of the report should be submitted at the end of the 9th week, and the final draft at the end of the 18th week upon feedback. Teaching of the fundamental theoretical knowledge and the research project are arranged as follows:

1.1 Lectures on Fundamental Theories

This course is to help the students grasp the basic concepts of the built environment and learn the fundamental knowledge of thermal and humidity environment, wind environment and air quality environment of the buildings and cities. By taking this course, the students can learn about theories such as the built environment and urban microclimate, come to understand human occupants' response to the thermal and humidity environment, know the actual application of the built environment and the latest research development, and lay a solid foundation for the measurement and simulation of building energy and the environmental system. The content of the course is as follows:

Built environment, urban physical environment, and solar radiation;

Thermal and humidity environment of the buildings, cooling load and heating load calculation principles and methods, and the thermal and humidity environment of cities;

Human occupants' response to steady-state and dynamic thermal and humidity environments and mathematical models of the response, indoor and outdoor thermal comfort, and thermal environment and productivity;

Natural ventilation and mechanical ventilation of the buildings, urban wind environment, and evaluation indicators of the indoor and outdoor wind environment;

Indoor air quality, urban air quality, impact of air quality on human and evaluation methods, and pollution control methods;

Basics of the acoustic and lighting environment of buildings and cities;

Building and urban environment measurement technologies.

1.2 Research Project

The "Built Environment" course is offered during the 2nd semester of the 1st year of the graduate program, which is an extremely important learning and transitional phase before the proposal of the master thesis. This course is not merely meant to impart the basic theoretical knowledge of the built environment to the students, but also lays a solid foundation for thesis writing by developing the students' research capacity. For this reason, we designed some research project topics based on the theories and research hotspots in this area for the students to select and conduct research projects (Table 1). The students can select a research topic according to their interest or thesis topic.

**2. Discussion of Training Practice**

"Built Environment" is an important basic course for the Building Environment and Equipment majors. The Building Technology major is rather new at Nanjing University, and most students majoring in this previously majored in Architectural Design. This poses a great challenge to education, but also brings an important opportunity to educational research. Spoon-fed education can get half the results with twice the effort, and the students' subjective initiative should be motivated. Therefore, this course attempts to apply an educational approach integrating the basic theoretical knowledge with the research project so as to motivate the students' learning initiative and develop their research capabilities.

2.1 Intensive Reading of the Textbooks

"Built Environment" explores how to develop the students' research capabilities; however, it should be noted that imparting basic theories is still the core teaching objective of this course. We cannot concentrate on details while forgetting the main objective. To ensure that the students master the basic theories, we apply intensive reading of the textbooks as the major teaching method. We also raise the following questions and similar questions in classes to improve the students' enthusiasm and effectiveness of intensive reading:

· Can we calculate the cooling load in summer with a steady-state algorithm of the daily mean temperature difference?

· What is the relationship between the ambient air temperature, mean radiant temperature and operating temperature?

· What is the relationship between the age of air, residual time and residence time?

......

The education on "Built Environment" is always committed to enhancing the students' comprehensive understanding of the comfort, health, energy saving and economic performance of the buildings and cities as well as their impact on the environment, and emphasize consideration for the relationships between indoor and outdoor, such as the indoor and outdoor evaluation indicators like comfort and ventilation, and the transport principles of indoor and outdoor pollutants.

2.2 Scientific Literature Retrieval and Reading

Research capability should be developed first upon mastering the content of the textbooks, and then by improving the literature reading capacity and applying a scientific and rational process. For example, if you want to study a green building technology, you need to learn its details, such as its application domestically and internationally, in various climate zones and in various areas, the earliest application, the most successful case, the failures and its main reasons, its impact on the environment, and current calculation method and existing software for technology evaluation from the textbooks and literature.

Literature retrieval and reading is a necessary capability for research work. Now, we will use the question "Do dust-floating floors exist" to illustrate how to improve the students' literature reading and analysis capability. In 2003, an article titled "A Female Realtor Tells You the Truth" appeared online. In the article, the sixth truth is that "You think the 9th to 11th floors are good ones? You are completely wrong, because they are dust-floating floors. When the dirty air arrives here, it would stop rising and stay here. But we would never tell the consumers about this." This statement may affect some consumers' purchase decision. Is it scientific? During literature retrieval, a student found a scientific paper whose abstract is "With probability theory, we set up a model for the widely concerned issue of the dust-floating floors. In so doing, we found out that the dust concentration rises with the height. Therefore, we proved the existence of the dust-floating floors", and completely accepted the conclusion. However, after further analysis, explanation and discussion in class, the students found many problems with the analysis method used in the article, and the conclusion is consequently questionable. Upon further retrieval, we found a newspaper article showing "A People's Daily report rebutted the statement 'the 9th to 11th floors are dust-floating floors' through field measurement, because the PM10 and PM2.5 concentration has nothing to do with the floors, but is related to the microclimate of the area where the building is located" (*People's Daily*, July 2, 2013, Page 4). Although this article is not a scientific paper, it employs a scientific analysis procedure; therefore, its conclusion is more reliable. From this example, the students realize that, not all of the scientific literature is reliable; they should judge the accuracy and reliability from various aspects and examine and screen the literature to carry out the research work in future.

2.3 Development of the Study Topic

Development of the students' research capability is a relatively long process that cannot solely rely upon the "Built Environment" course. However, the design and implementation of the research project in this course can positively help to develop students' research capabilities. For example, during the process from topic selection to determination of the midterm project topic, and then to determination of the final project topic, most students evolve their topics from a general and conceptualized form to a specific and operable form, except the few ones who always keep their topics unchanged. In addition, some students may abandon the original topic for various reasons (Table 2). These are common issues occurring in research, and earlier encounter can help to lay a foundation for their future research work.

2.4 Training Achievement Demonstration

After a semester of learning and specialized study in the "Built Environment" course, most students mastered the skills of retrieving Chinese and English scientific literature based on the relevant topic and applying the theoretical knowledge of Built Environments to the research project and architectural design, except for a few students who still have no conception of research. The Figure 1 shows the results of an analysis on a specific built environment that a student obtained upon literature search, and Figure 2 shows another student's built environment design with consideration of the local environmental factors. Although the students' research project is still at a superficial stage due to the limitation of the class hour, some initial results have been achieved. In future, one of the projects deserving our consideration is how to trace the students to make sure whether they can apply the knowledge and study tools to the research activities.

Table 1 List of the Opening Study Topics for the Course of "Building Environment"

| Building Environment and Energy Saving Technology | Wind Environment and Air Quality | Urban Microclimate | Human Bodies' Response to the Thermal and Humidity Environment |
|---|---|---|---|
| Solar Integrated Technology | Indicators of Ventilation Efficiency: From Indoor to Outdoor | Urban Microclimate Simulation Software | Evaluation of the Indoor and Outdoor Thermal Comfort Indicators |
| Solar Trombe Wall | Comparison of the National and Industrial Standards on Indoor and Outdoor Air Quality | Calculation Area When Using CFD for Outdoor Environment Calculation | Thermal Environment and Labor Efficiency |
| Solar Chimney | | | |
| Double-Skin Facade | The Transmission Relationship between the Indoor and Outdoor Pollutants (Suspended Particles and Ozone) | Analysis on the Accuracy of the CFD Simulation Technology | — |
| Low-e Glass | Study on the Bioaerosol Features | Plant, Tree Crown and Vehicle Simulation Method | — |
| Photovoltaic Technology | Technology of Sterilization by the Air Purifier | Urban Microclimate Simulation by the Porous Media Method | — |
| Vertical Greening and Roof Greening Technology | Efficiency of the Air Purifier | — | — |
| Rainwater Recovery System | UV Light Sterilization Technology | — | — |
| Environmental Assessment on Temporary Architecture such as Olympic Venues and Expo Sites | Effect of Green Plants on Purifying the Air | — | — |

Table 2 Development Process of the Study Topics of "Building Environment"

| Topic | Midterm Dissertation Topic | Final Dissertation Topic |
|---|---|---|
| Solar and Building Integration Technology | Trombe Wall | Reasonable Dimensions of the Traditional Trombe Wall and Construction Method and Applications |
| Solar and Building Integration Technology | Double-Skin Glass Facade | Study Progress of the Double-Skin Facade System |
| Analysis and Study of the "Dust-floating Floors" | Double-Skin Facade | Think about the Double-Skin Facade Technology and Local Practice Design |
| Roof Garden Literature Review | The Effect of Roof Greening on Indoor and Urban Environment Improvement | Discussion on Roof Greening in a Climate Zone with Hot Summers and Cold Winters |
| Vertical Greening and Facade Design | The Ecological Function of Vertical Greening | Pest Control in Vertical Greening |
| Mathematical Modeling of Plants | A Review of the Research on Thermal Environment Simulation of Trees in the Urban Microclimate | A Summary of the Specific Methods used for Evaluating the Effect of Plants on the Outdoor Microclimate with Computer Simulation Technology |
| Building Ventilation | Criteria for Indoor Ventilation Evaluation and How to apply the Evaluation Indicators for Indoor Ventilation to Outdoor Ventilation | Evaluation Indicators for the Airflow Organization and Analysis and Study on Applying the Evaluation Indicators for Indoor Ventilation to Outdoor Ventilation |
| Indoor Pollutant: Ozone | Efficiency and Damage of UV Light Sterilization | Efficiency and Damage of UV Light Sterilization |
| Plants and Indoor Air Pollution | Application of Plants to Indoor Formaldehyde Elimination | Progress of Study on Indoor Formaldehyde Elimination by Green Plants |
| Indoor Air Pollution and Countermeasures | Indoor Air Purifier | Negative Ion Indoor Air Purifier |
| Study on Architectural Design of Silkworm Farms | Analysis and Study on Architectural Design of Silkworm Farms | Analysis and Study on Architectural Design of Silkworm Farms |

年度改进课程
WHAT'S NEW

# 设计基础
## BASIC DESIGN
鲁安东 丁沃沃

**动作装置**

  设计一个装置，作用于身体动作并改变人与环境的关系。目的在于使学生初步认识身体、尺度与环境的相互影响；学会观察并理解场地；初步认识形式与背后规则的关系；学会发现日常使用中的问题并解决问题；学会使用分析图交流构思。

**折纸空间**

  利用折叠纸板创造一个多层空间，并将其转化为居住空间。目的在于使学生初步掌握二维到三维的转化，初步认识图和空间的再现关系；利用单一材料围合复杂空间,初步认识结构与空间的关系；学会用分析图进行表述。

**覆盖结构**

  建造一个覆盖大尺度空间的结构，作展示用途。目的在于初步理解支撑体系和围护体系；初步建立造价概念；初步理解力的关系，学会感受结构美和空间美；初步认识节点设计；强化场地意识（包括朝向、环境、流线等）；学会用分析图进行表述。

**Act installation**

Design an installation that acts on bodily actions and redefines human-environment relationship. Its purposes are to inform the students of the mutual influences between body, scale and environment; to learn to observe and interpret a situation; to formulate an understanding of form and its underlying principles; to learn to identify problems in everyday use and conduct problem-solving; and to learn to communicate ideas with analytical drawings.

**Folding space**

Fold a cardboard to create a multi-layered space and dwell that space. Its purposes are to enable the students to translate between 2D and 3D to understand the representational relationship between drawings and space; to enclose complex space with one material to understand the relationship between structure and space; and to learn to communicate ideas with analytical drawings.

**Covering structure**

Build a structure that covers a real-scale space for exhibition. Its purposes are to introduce a fundamental understanding of the supporting and enclosure systems and of the budget; to understand the roles of structural forces and the beauty of structure and space; to learn to design the joints and establish an understanding of site (including orientation, ambience and circulation, etc.); and to learn to communicate ideas with analytical drawings.

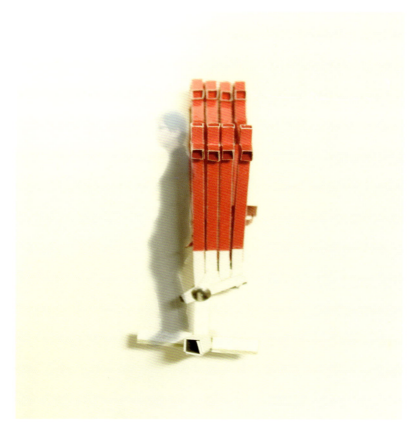

根据身体动作的运动轨迹生成的装置：它依次展开的外轮廓模拟了身体运动所需的最小空间。
An installation based upon the traces of bodily movement: its unfolding profile captures the minimum space required for the performance of an action.

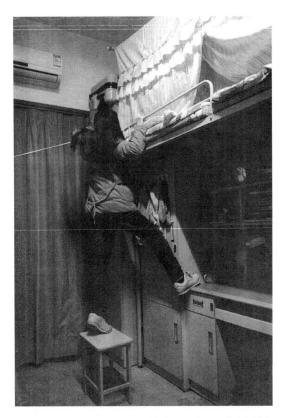

观察日常动作"爬床";分析手脚的配合与空间的限定关系;针对该动作设计联动装置;对该装置的作用原理进行分析图示。
Observe an everyday action: climbing up to bed; analyze the synchronization of the movements of arms and legs and their relationship to the surrounding space; design an installation that connects the movements of the arms and legs; use diagrams to visualize the rationale of this installation.

 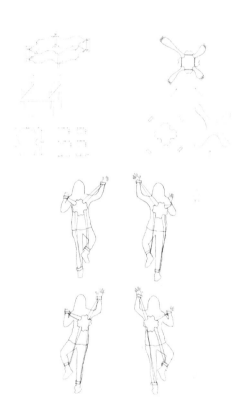

观察日常动作"压腿";分析压腿动作中肢体的运动轨迹;针对该动作设计的装置,便于抬腿和下腰;对该装置的几何与尺寸进行分析图示。
Observe an everyday action: a stretching exercise; analyze the traces of body movement in this action; design an installation that facilitates the body movement; use diagrams to visualize its geometry and dimensions.

 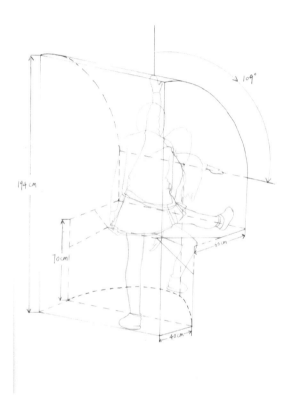

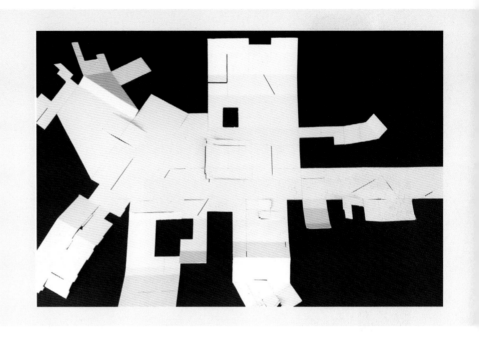
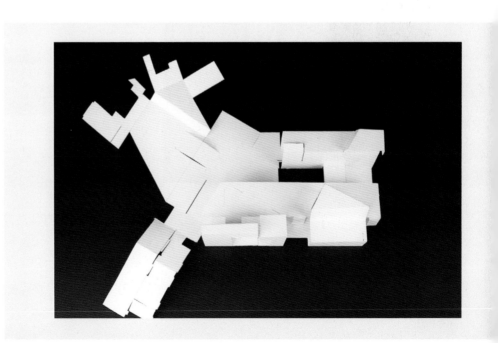

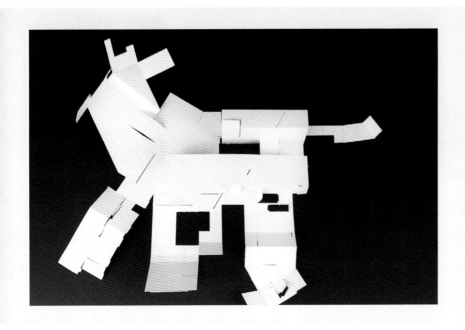

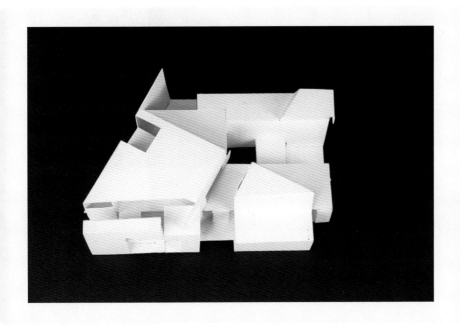

从一张 A1 卡纸出发，分四步进行一系列的折叠和插接，组合成有趣的空间。本方案通过引入斜线形成了三角形与长方形体量的交错。
Start with an A1-size cardboard and, in four steps, use a series of folding operations to create interesting spaces. This project has introduced diagonal lines to formulate a juxtaposition of triangular and square volumes.

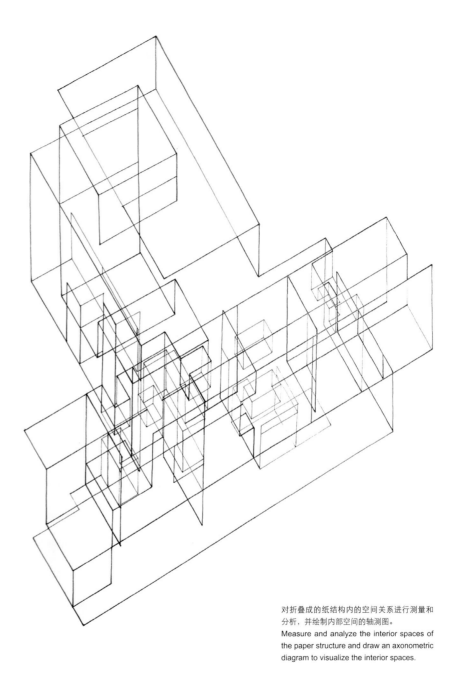

对折叠成的纸结构内的空间关系进行测量和分析,并绘制内部空间的轴测图。
Measure and analyze the interior spaces of the paper structure and draw an axonometric diagram to visualize the interior spaces.

选择纸盒内四个有趣的房间,用不同色彩和材质的纸拼贴出这些房间,并用拼贴的形式将这四个房间组织为一个连续的空间漫步。
Choose four interesting rooms inside the paper structure and use paper of different colors and textures to visualize these room. Then use collage method to connect these rooms into a continuous spatial promenade.

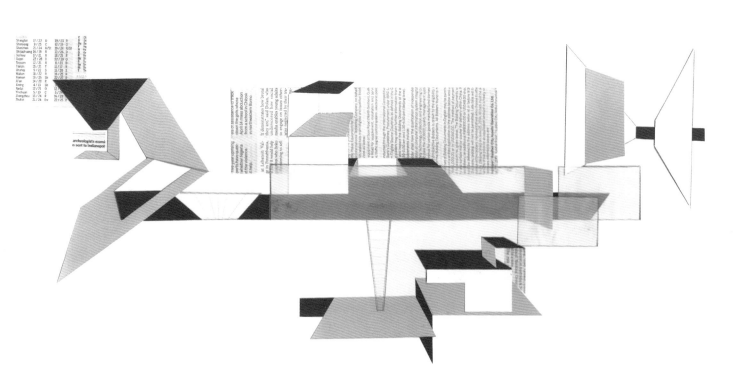

这一方案通过三种材料单元，拼合成二种结构构件，进而组成一个供人休息的结构体。设计逻辑清晰，但创造了丰富的视觉形态。
This project has used three types of units to fabricate two kinds of structural components and used them to assemble a structure that shelters people. It has a clear design logic but created diverse visual forms.

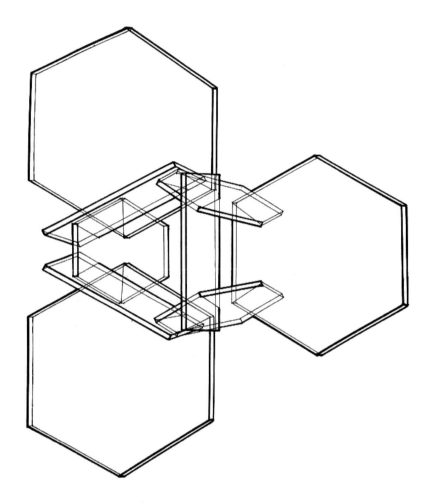

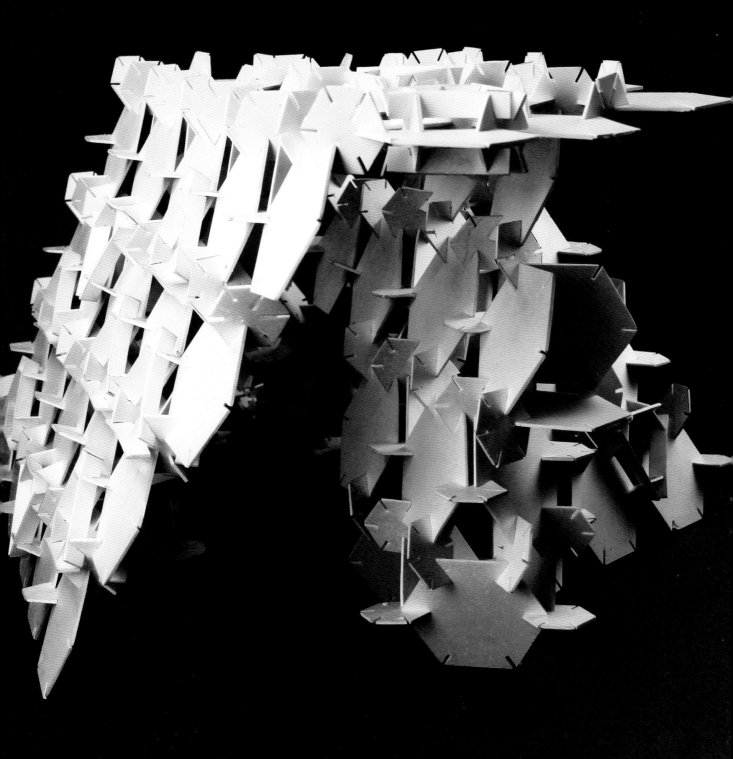

建筑设计（一）ARCHITECTURAL DESIGN 1
# 风景区茶室设计
TEA HOUSE DESIGN

刘铨 冷天 王铠

　　本练习的基地选择学生在环境认知练习中调研过的城市风景区内两处临水的坡地。建筑功能为茶室，包括可供至少 60 人使用的大厅和 30 人使用的雅座，以及操作间、休息室、洗手间等必要的辅助用房，并满足无障碍设计的要求，总建筑面积不超过 300 ㎡，时间 7 周。

　　第一周：制作 1：200 纸质模型帮助研究分析场地地形条件，主要是景观朝向与场地标高变化，多方案构思空间并以模型表达。

　　第二周：通过 A3 图幅 1：200 比例的平面与剖面草图，在确定方案的基础上调整标高、空间划分、功能流线与场地的关系。

　　第三周：制作 1：100 纸质模型和 SketchUp 模型来帮助进行建筑空间与场地关系的进一步深化调整。

　　第四周：通过 A3 图幅 1：100 比例的平面与剖面草图，根据功能要求在剖面标高变化、尺度和结构方面对空间进行细化。

　　第五周：使用透视图和模型照片表达空间设计意图，通过模型的透视角度，研究立面材料与构造，完成立面与墙身节点的设计。

　　第六周与第七周：正式成果图纸、模型的制作，认知排版作为设计的一部分，它与设计意图、设计过程表达的关系。

This section is to design a tea house on the slope near the lake. The tea house includes a hall for 60 persons, several private rooms for 30 persons and other necessary service space. The accessibility design of the building is required. The site of this project is also chosen from the areas which the students investigated in the last semester.

1st week: make 1:200 physical models to support the analyze of site topography conditions, mainly the view point and site loops, to find out multiple proposals of space.

2nd week: by 1：200 plan and section sketches, to adjust relations of site and space division and function flow.

3rd week: make 1：100 physical model and SketchUp model to support further adjustment on relation of spaces and the site.

4th week: by 1：100 A3 plan and section sketchs, draw details of space elevation changes, dimensions and structures according to function requirements.

5th week: interpret space design concepts via perspectives, and choose building facade materials and structures from the perspective point of view. Draw 1：20 wall details to accomplish design of the building.

6th & 7th weeks: make official rendering picture and model, and understand relations of the layout of drawings, as part of design, to design concepts and design process.

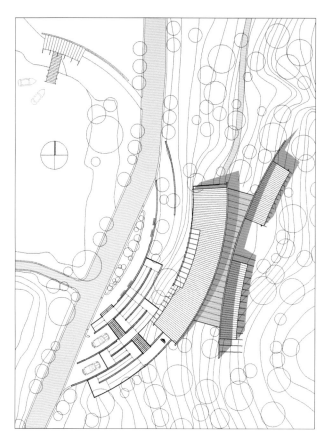

该方案使用与地形等高线相平行的系列弧墙来契合场地的坡面,并形成了三个标高的长条形空间来应对不同的景观朝向。室外场地的设计也采用了一致的空间形式语言,把坡道、踏步停车场有机地组织到地形环境中。

The scheme uses a series of arc walls parallel with the topographic contour to fit into the slope surface of the site and forms the rectangular space of three elevations to cope with different landscape orientation. The design of the outdoor site also adopts consistent spatial form language to organize the ramp and the parking lot into the terrain environment organically.

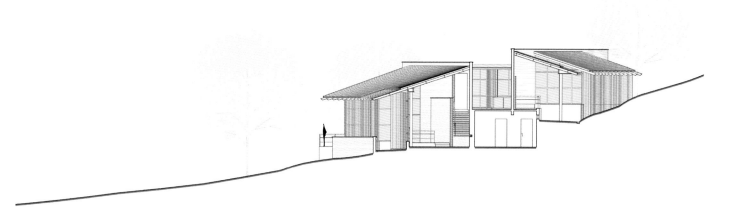

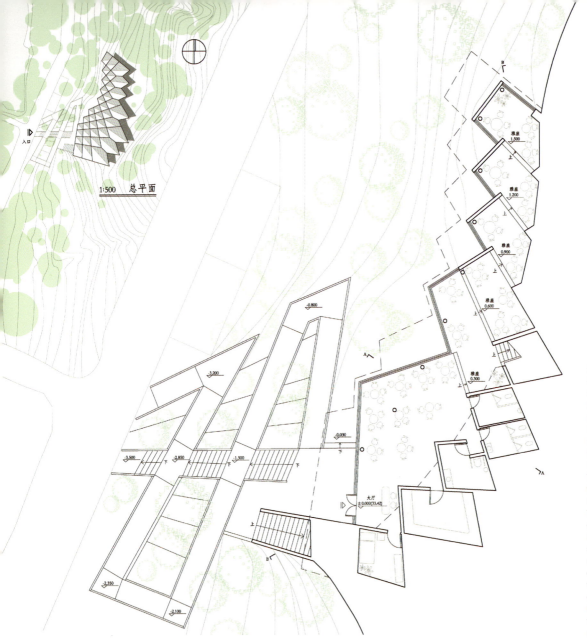

这一方案则是由两个在不同标高上的平行等高线布局的体量发展而来，通过平面网格的设定，将较大体量的垂直界面挤压为锯齿状，既能更好地融合在自然环境之中，又有利于内部空间的再划分和景观视线的引导，屋顶也在此网格基础上进行了设计，树叶形的单元不仅与环境相呼应，也发挥出采光和通风的作用。

The scheme is developed from the volume of two parallel contour line layouts at different elevations. Through the setting of the planar gridding, the larger volume vertical interface is extruded into the zigzag shape, which can better fit into the natural environment and help re-divide the internal space and guide the landscape sight. The roof is also designed on the basis of gridding. The leaf-shaped unit not only echoes with the environment but also plays the role of lighting and ventilation.

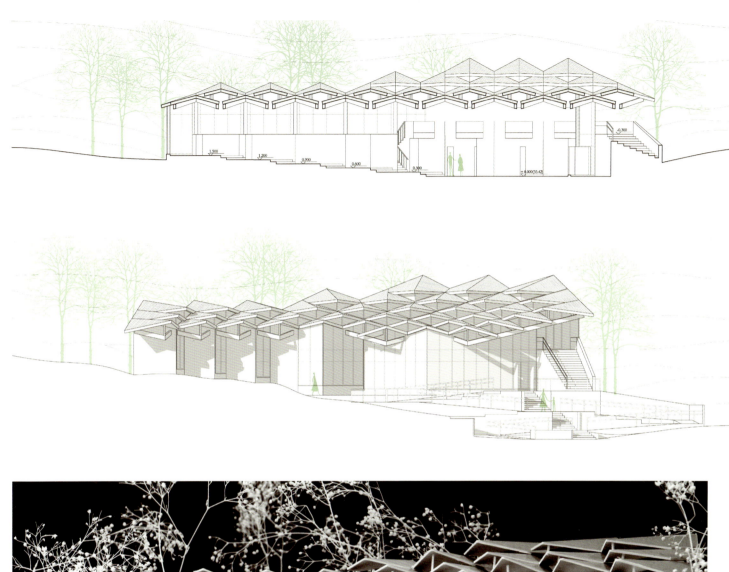

该方案在平行与垂直于地形等高线体量的基础上设计了连续的折板，不仅引导了流线，也框定了景观视野。
The scheme designs continuous folded plate on the basis of paralleling and perpendicular to the topographic contour volume, which not only guides the flow line but also frames the landscape view.

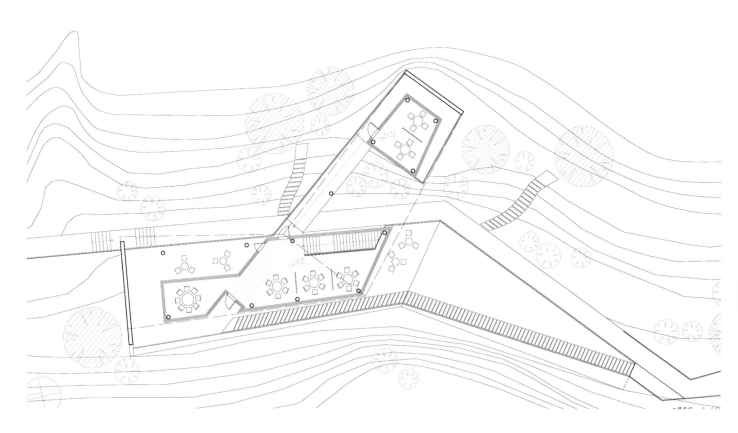

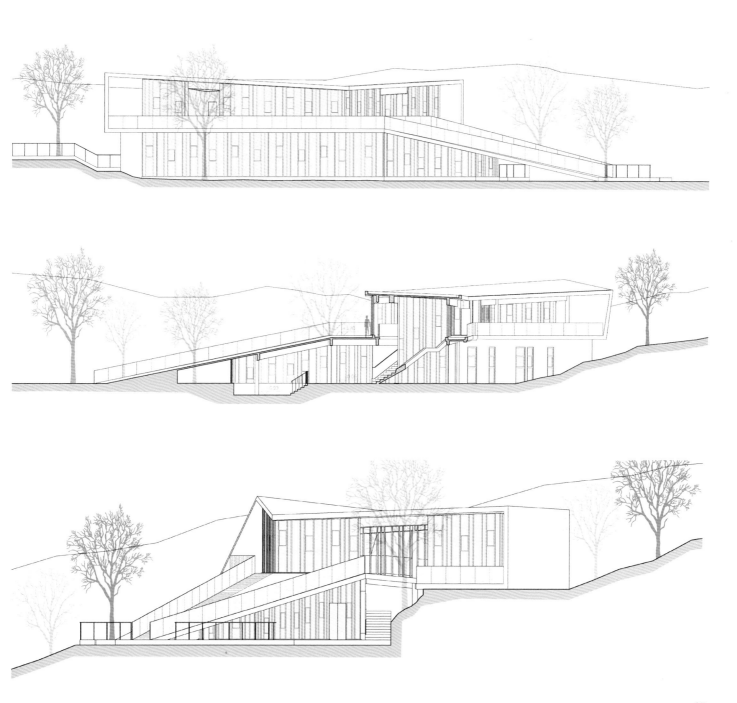

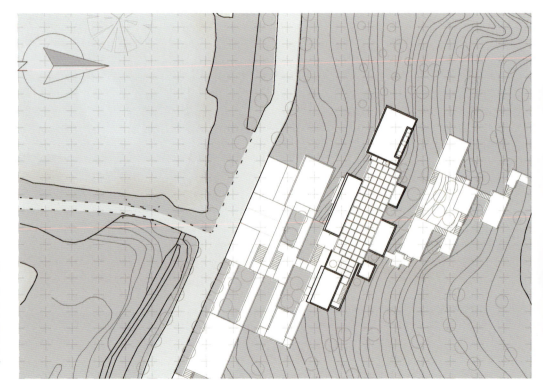

该方案以控制网格为基础，设计了众多相互并置、咬合、嵌套的盒体，以减小体量、适应场地坡面及限定室内景观视线方向。
The scheme designs many mutually juxtaposed, occlusive and nested boxes on the basis of controlling gridding to reduce the volume, adapt to the site slope and limit the indoor landscape gazing direction.

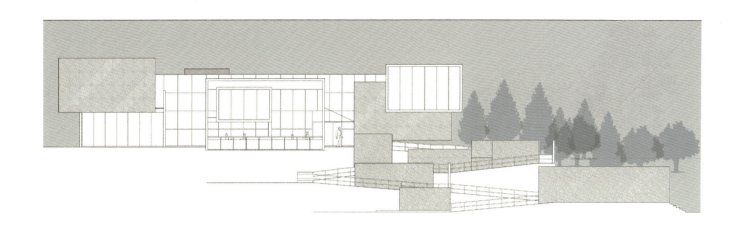
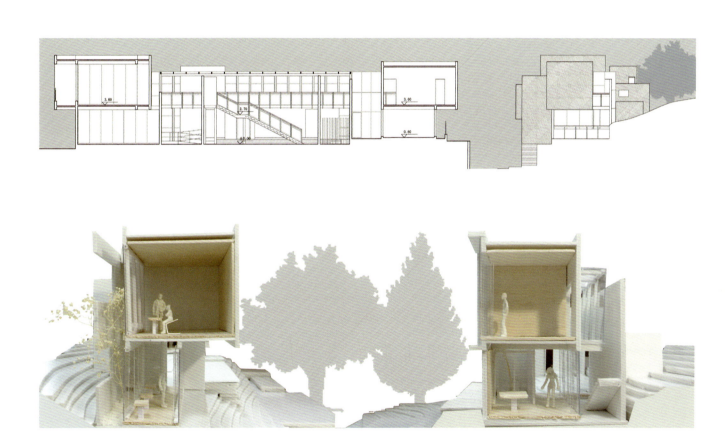

39

建筑设计（二）ARCHITECTURAL DESIGN 2

# 赛珍珠纪念馆扩建
## EXPANSINO OF PEARL BUCK'S HOUSE

周凌 童滋雨 钟华颖

**教学目标**

　　此课程训练最基本的建造问题，使学生在学习设计的初始阶段就知道房子如何造起来，深入认识形成建筑的基本条件：结构、材料、构造原理及其应用方法，同时课程也面对场地、环境和功能问题。训练核心是结构、材料、场地。在学习组织功能与场地同时，强化认识建筑结构、建筑构件、建筑围护等实体要素。

　　文脉：充分考虑校园环境、历史建筑、校园围墙以及现有绿化，需与环境取得良好关系。

　　退让：建筑基底与投影不可超出红线范围。若与主体或相邻建筑连接，需满足防火规范。

　　边界：建筑与环境之间的界面协调，各户之间界面协调。基底分隔物（围墙或绿化等）不超出用地红线。

　　户外空间：扩建部分保持一定的户外空间，户外空间可在地下。

　　地下空间：充分利用地下空间。

**任务书**

　　基地内地面最大可建面积约100 $m^2$，地下可建面积200—300 $m^2$，总建筑面积约300—400 $m^2$，建筑地上1层，限高4 m，地下层数层高不限。展示区域 200—300 $m^2$，导游处10 $m^2$，纪念品部30 $m^2$，茶餐厅60 $m^2$，厨房区域大于10 $m^2$，门厅与交通，卫生间。

**Training Objective**

This course trains the students to solve the basic construction of architecture. Students should learn how to build an architecture at the very beginning of their studying, understand the basic aspects of architectures: the principles and applications of structure, material and construction. The course also includes the problem of site, enviroment and function. The keypoints of the course include site, structure and material. Students should strengthen the understanding of physical elements including structures, components and façades while learning to organize the function and site.

Context: The enviroment, historical building, the edge of the campus and the green belt around the site should be taken into consideration. The expansion is expected to have a good relationship with the surroudings.

Retreat Distance: The new architecture can't beyond the red line. Fire protection rule should be complied.

Boundary: Both the boundary between different buildings and between building and environment should be harmonized.

Open Space: Open space should be considered, which permitted to be placed underground.

Underground Space: Underground space should be well used.

**Mission Statement**

The maximum ground can be used in the base area is about 100 $m^2$, while underground construction area is about 200-300 $m^2$, and the total floor area of architecture should be about 300-400$m^2$.The architecture should be 1floor above the groud lower than 4m.The underground levels have no limitation. Exhibition area: 200-300$m^2$, Information center: 10$m^2$, Shop: 30$m^2$, Coffee bar: 60$m^2$, Kitchen: >10$m^2$, Lobby & walking space, Toilet .

采用双坡"盒中盒"的套叠关系,巧妙处理新建筑与南北场地的关系。
Adopt the double-way gradient "box-in-box" nested relation and smartly deal with the relationship between new buildings and the site from north to south.

通过内外盒的材质、位置以及联系方式的变化，内盒、外盒与旧馆间形成多重行为和视觉关系。
The inner box, outer box and the old pavilion form multiple behavior and visual relationships through the changes of the materials, positions and contact information of the inner and outer boxes.

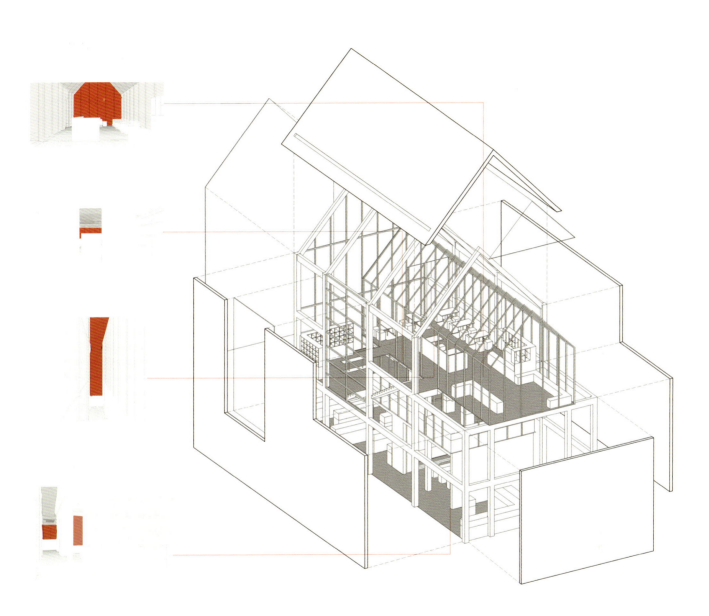

内外盒主要构件关系分解轴测
The isometric perspective of the main component relationship of inner and outer boxes

以三角形为母题，既与原环境有易读的文脉呼应，又限定出建筑自身在外部形态、内部空间和结构体系上的统一。
With the triangle as the motif, it not only echoes with the readable context of original environment but also restricts the unification of buildings in the external form, internal space and structural system.

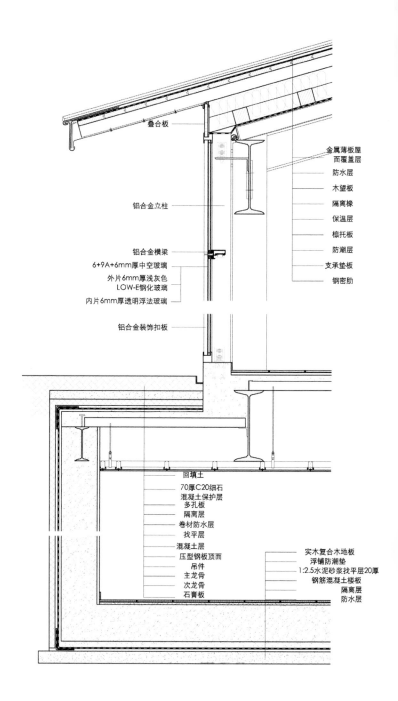

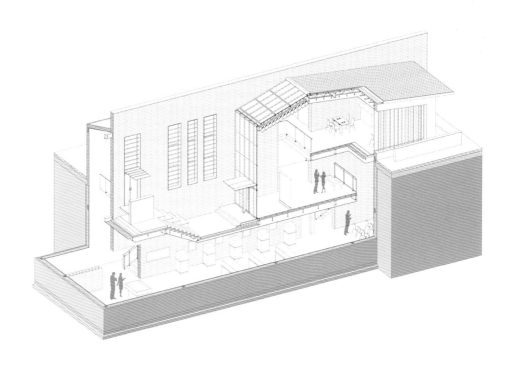

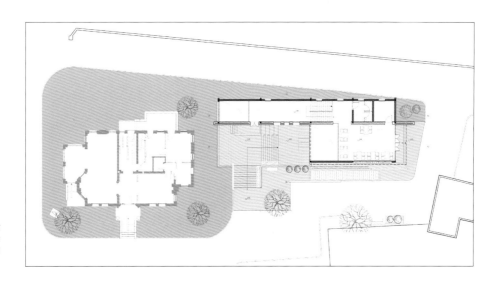

以旧馆外墙的延伸线作为新建筑的分区依据，发展出关系清晰层次丰富的空间体系。
The extension line of the old pavilion exterior wall is taken as the partition basis of new buildings to develop the three-dimensional system with clear relationship and rich layers.

下部片墙形成清晰的流线和视线导向，上部剖屋顶和旧馆的关系虽有文脉争议但也不乏空间张力。

The lower wall forms clear flow line and sight guiding. The relationship between the upper roof and old pavilion has the context controversy but does not lack spatial tension.

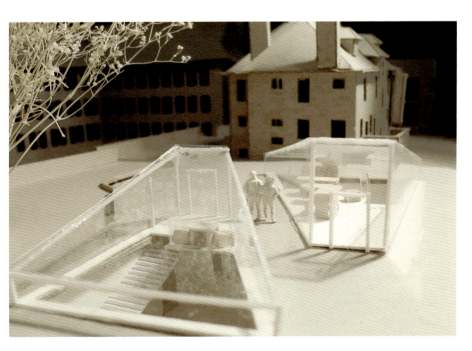

贯穿场地的通道将地面以上隔成形体相似的两部分,创造出更积极的新旧建筑场所关系。
The passageway running through the site divides the building above the ground into two parts with similar shapes and creates a more active old and new building relationship.

建筑设计（三）ARCHITECTURAL DESIGN 3
# 大学生活动中心设计
## DESIGN OF THE COLLEGE STUDENT CENTER

周凌 童滋雨 钟华颖

**教学目标**

课程主题是"空间"，学习建筑空间组织的技巧和方法，训练空间的效果与表达。空间问题是建筑学的基本问题，课题基于复杂空间组织的训练和学习。从空间秩序入手，安排大空间与小空间、独立空间与重复空间，区分公共与私密空间、服务与被服务空间、开放与封闭空间。训练重点是空间组织，包括空间的秩序、空间的内与外、空间的质感及其构成等。以模型为手段，辅助推敲。设计分阶段体积、空间、结构、围合等，最终形成一个完整的设计。

**任务书**

1. 空间组织原则

空间组织要有明确特征，有明确意图，概念要清楚，并且满足功能合理、环境协调、流线便捷的要求。注意三种空间：聚散空间（门厅、出入口、走廊）；序列空间（单元空间）；贯通空间（平面和剖面上均需要贯通，内外贯通、左右前后贯通、上下贯通）。

2. 空间类型

多功能空间、展示空间、专属空间、休闲空间、服务空间、交通空间。

总建筑面积控制在 3000 $m^2$ 以内，层数控制在 4 层以内。

**Training Objective**

The theme of the course is space. It trains students' skills of space organization and make them master methods of space organization and presentation. Space issues are the basic issues of architecture. This course organizes trainings and studies based on complex space. Students start with spatial order, arrange the large spaces and small spaces, independent spaces and repetitive spaces, differentiate public and private spaces, service and serviced spaces, open and close spaces. The key point of training is space organization, including the order of space, inner and outer, texture of space and its composition, etc. Models should be used as means to assisting deliberation. The course includes stages of volume, space, structure, enclosing, then forms a complete design.

**Mission Statement**

1. Principle of space organization

The space organization should have clear characteristics, intentions and concepts, while satisfying the requirements for reasonability, environmental harmony and convenient circulation. Gathering and dispersing spaces (hallway, entrance / exit, corridor); Sequence space (unit spaces); Through space (on plans and sections, including internal/external through, left / right / front / rear through, upper/lower through).

2. Type of space

Multi-funtion space,exhibition space,proprietary space,recreation space,service space,circulation space: lobby, corridors, etc.

The total floor area ≤ 3,000 $m^2$, ≤ 4 floors.

以"管道"为元素，限定出穿插交织的互反两类空间，创造丰富、纯粹、有趣的空间。
Limit two interlaced reciprocal spaces with "pipelines" as the element and create a rich, pure and fun space.

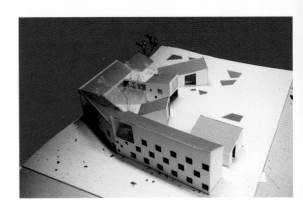

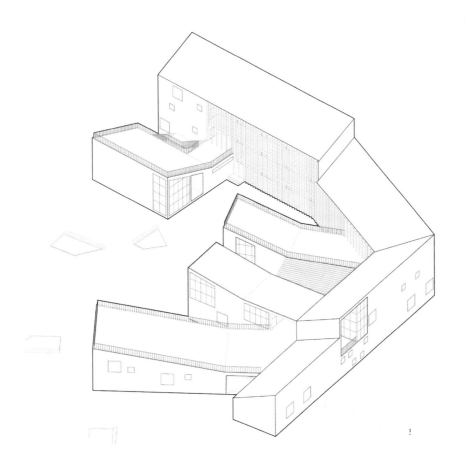

一条连续的折"线"与多个散落的"点"回应了不同的边界条件,引发出多样的活动模式。
A continuous folding "line" and many scattering "points" respond to different boundary conditions and arouse multiple activity modes.

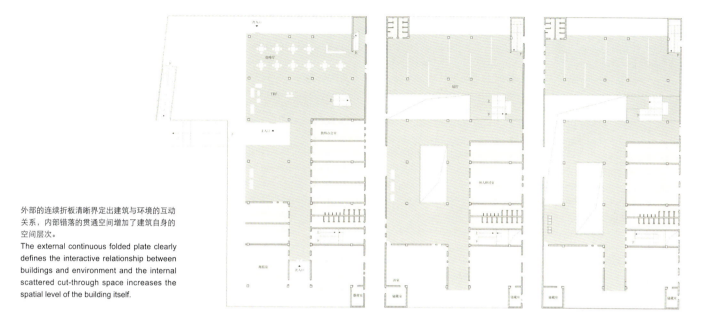

外部的连续折板清晰界定出建筑与环境的互动关系，内部错落的贯通空间增加了建筑自身的空间层次。
The external continuous folded plate clearly defines the interactive relationship between buildings and environment and the internal scattered cut-through space increases the spatial level of the building itself.

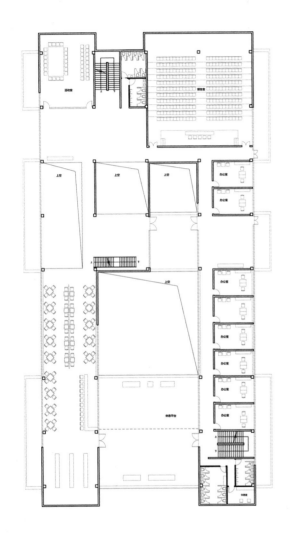
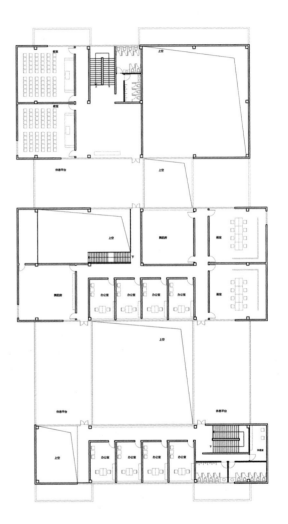

通过积木的概念化整为零，把大空间拆分成三层积木纵横交叠起来形成基本形体，用简单的操作活动丰富的空间以及合理的功能。
Break up the whole into parts by the concept of building blocks, split the large space into three layers of building blocks overlapped and use simple operations to activate rich spaces and reasonable functions.

均质网格与变化的空间。
Create varied spaces in homogeneous gridding.

建筑设计（四+五）ARCHITECTURAL DESIGN 4 & 5
# 社区商业中心+观演中心
## COMMUNITY BUSSINESS CENTER & PERFORMANCE CENTER

华晓宁 王丹丹 钟华颖

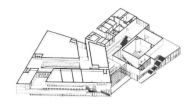

**教学目标**
　　两个课题的设计内容被设定为同一城市场址上整体发展计划的不同组成部分。学生在一开始就必须将注意力集中于场地与城市的关系研究，通过两个新建筑的介入，回应既有城市文脉，并在场址上生成新的城市关系和城市物质空间系统。在方案发展过程中，学生也必须始终保持多种视角的转换，同时考察建筑自身、建筑与建筑之间、建筑群体与城市物质空间系统之间复杂的互动关系。因城市而生成建筑，同时也因建筑而生成城市。

**研究主题**
　　实与空：关注城市中建筑实体与空间的相互定义、相互显现，将以往习惯上对于建筑本体的过度关注拓展到对于"之间"（in-between）的空间的关注。
　　内与外：进一步突破"自身"与"他者"之间的界限，将个体建筑的空间与城市空间视为一个连续统，建筑空间即城市空间的延续，城市空间亦即建筑空间的拓展，两者时刻在对话、互动和融合。
　　层与流：不同类型的人和物的行为与流动是所有城市与建筑空间的基本框架，当代大都市中不同的流线在不同的高度上层叠交织，构成一个复杂的多维城市。必须首先关注行为和流线的组织，由此才生发出空间的系统和形态。
　　轴与界：城市纷繁复杂的形态表象之后隐含着秩序和控制性，并将成为新的形态介入。

**Training Objective**
The design objects of two studios are set as two integral parts of a single development program on one urban site. Students should focus on the relationship between site and city from very beginning, respond urban context by the intervention of two new architectures, then generate the new urban relationship and urban physical space system. Further, students should keep on drifting viewpoints throughout the development of projects, studying the complex interactions between different architectures and urban physical space system. Architectures emerge from city, while city emerges via architectures.

**Research Subject**
Volume &Void:Focus on the inter-definition and inter-generation between the volume and void in the city. Expand the attention from the architecture itself to the in-between space.
Inner & Exterior:Break the boundary between the inner and exterior, regard the architectural space and urban space as a continuum. Architectural space is the continuation of urban space, while urban space is the expansion of architectural space. Both keep on dialoguing, interacting and fusing.
Layer&Circulation:Behaviors and circulations of different persons and things are basic frames of all architectural and urban space. Incontemporary metropolis, various circulations overlap and interweave on various layers, thus form a miscellaneous multi-dimension city. The organization of behaviors and circulations should be studied first, from which the system and form of space could be generated.
Axis & Edge:There are orders behind the heterogeneous imageries of urban form, which should be the alignments of new interventions.

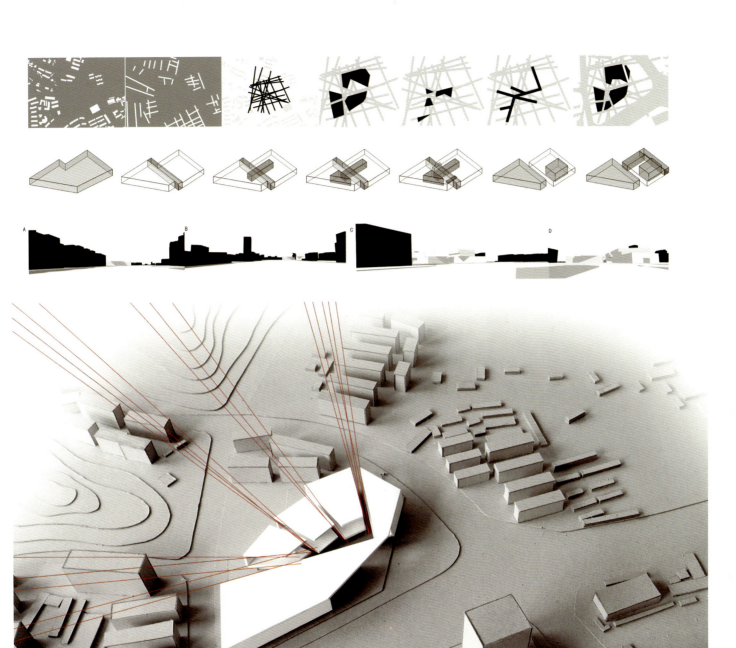

一阶段：场地与城市关系分析、出入口布置、场地流线初步组织建筑形体与外部空间的初步布局、初步的功能分化。
Phase 1:analyzes of site and urban relationship, organization of exits and circulations, arrangement of volume and void/ configuration of function and program.

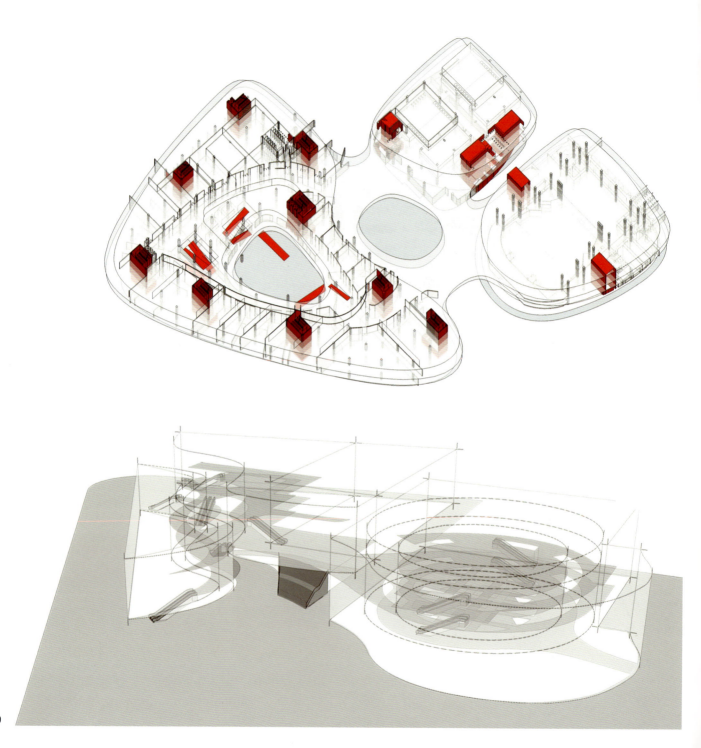

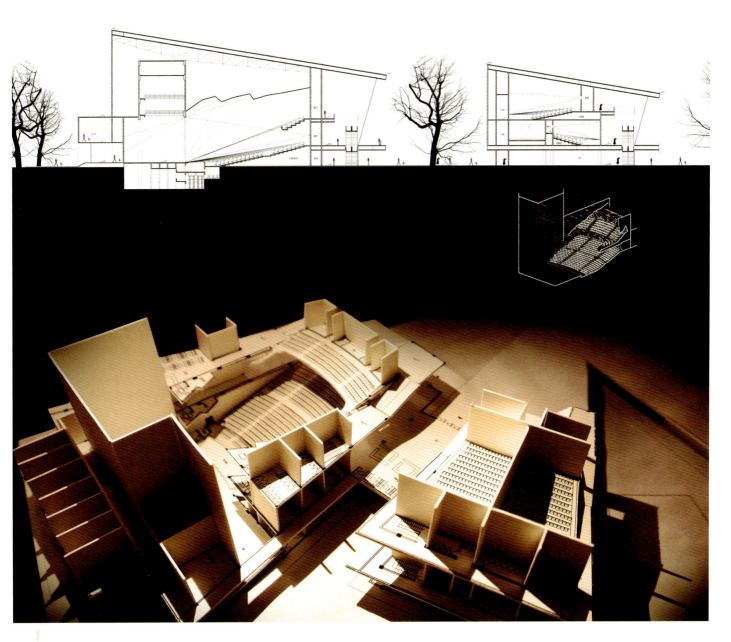

第二阶段：整体公共空间系统的建构、观演厅堂的深化设计。
Phase 2: construction of the public space system throughout two architectures, deepin design of performance space.

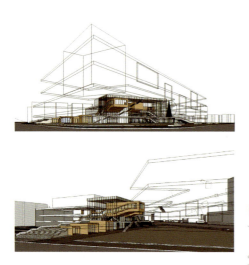

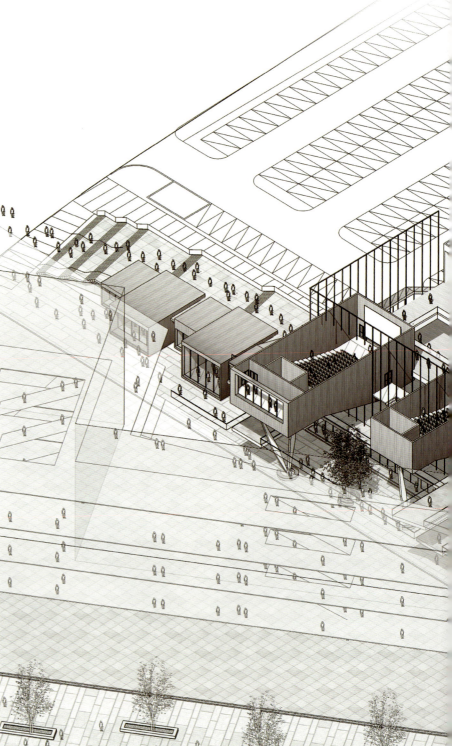

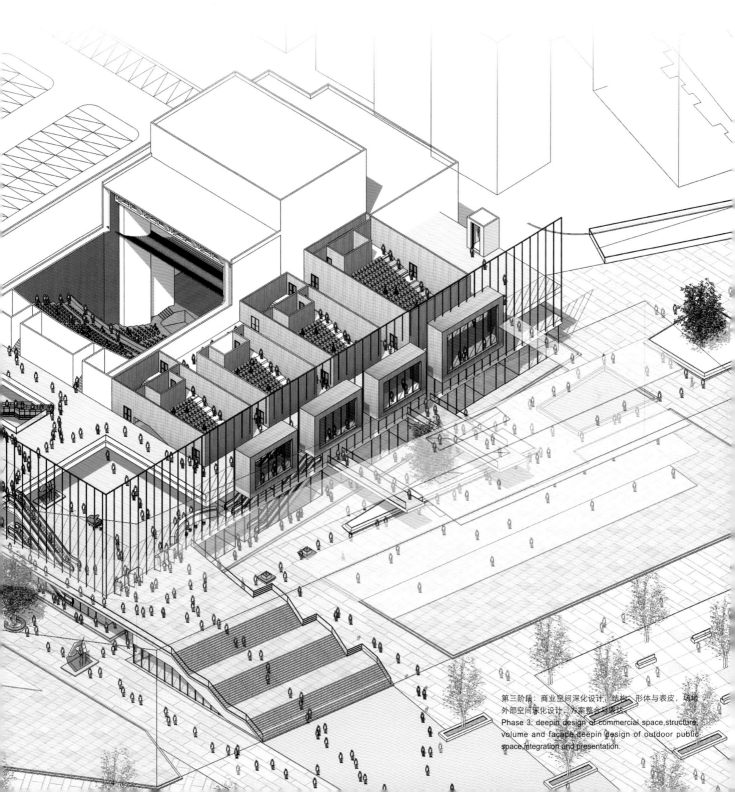

第三阶段：商业空间深化设计，结构、形体与表皮，场地外部空间深化设计，方案整合与表达。
Phase 3: deepin design of commercial space, structure volume and façade, deepin design of outdoor public space, integration and presentation.

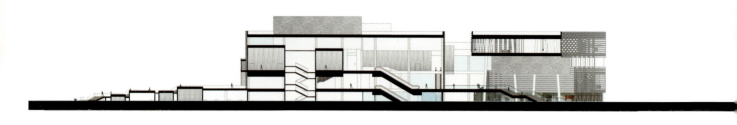

本科毕业设计 GRADUATION PROJECT

# 洪家大屋测绘与改造设计
## MAPPING AND TRANSFORMATION DESIGN OF YANSHE

周凌 胡友培

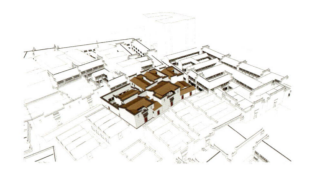

**教学目标**

中国传统村镇正在消失，其速度与中国城市化速度成正比，传统农耕文化、传统手工艺、传统价值观处于消失弱化的边缘。本课题着重研究中国传统村镇保护更新的议题，以安徽祁门洪家大屋及周边地块保护更新规划设计为例，对传统村镇进行基础研究，在研究基础上，提出整体保护和更新的概念规划。学生通过调研和规划设计，了解传统村镇与传统建筑文化，学习规划知识，训练建筑设计技巧。

**研究主题**

村落自然与人文环境、民居类型、保护规划、改造策划、技术与建造。

**作业要求**

通过实地调研，掌握古建筑知识与制图技巧；通过案例研究学习方法，通过讨论梳理思路，最终掌握规划与设计中分析图、平立剖图、模型等表达，完成规划与建筑设计深度的图纸与研究报告。图纸表现方式和比例自定。

**Training Objective**

Chinese traditional villages and towns are fading away in direct proportion to Chinese urbanization speed, and the traditional farming culture, traditional handicraft and traditional values are on the verge of disappearing and weakening. The course focuses on studying the topic about Chinese traditional village and town protection and updating. Taking Hong Family's big house in Qimen, Anhui and surrounding land protection and updating plan for example, it makes basic research on traditional villages and towns and puts forward the concept planning of overall protection and updating on the research basis. Students understand the traditional villages and towns and traditional building culture through survey and planning design, learn planning knowledge and practice building design skills.

**Research Subject**

Context and culture，typology of traditional building，master planning for preserving，program and function，technique and construction．

**Assignment Requirement**

Master the historic building knowledge and drawing skills through field survey; study learning methods in cases, sort thoughts in discussion, finally master the analysis graphics, plane, vertical and cross-sectional drawings, models and other expressions in planning and design, and complete the drawing and research report of planning and building design depth. The drawing manifestation mode and proportion are decided for themselves.

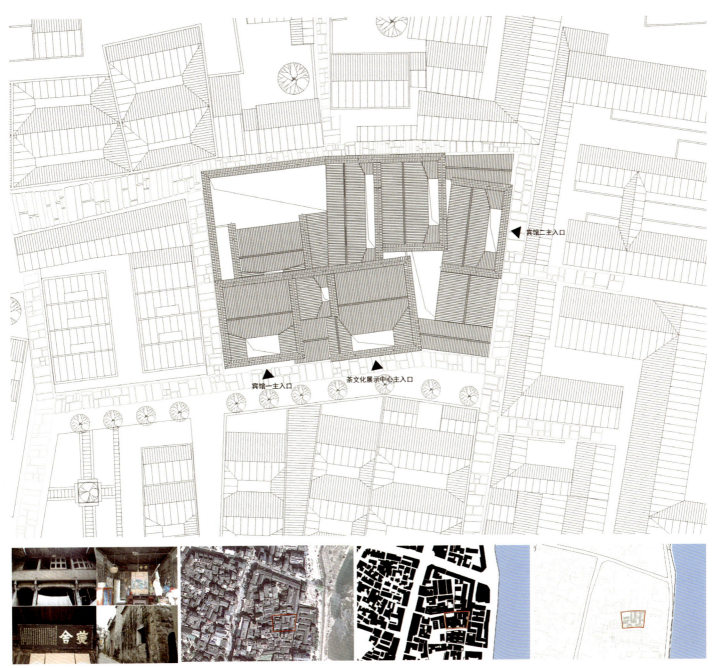

基于测绘数据，将兼具徽派建筑和民国建筑风格的燕舍改建成茶馆和宾馆。
Rebuild the Yan House with Hui style architecture and the architectural style in the Republic of China into teahouses and hotels on the basis of surveying and mapping data.

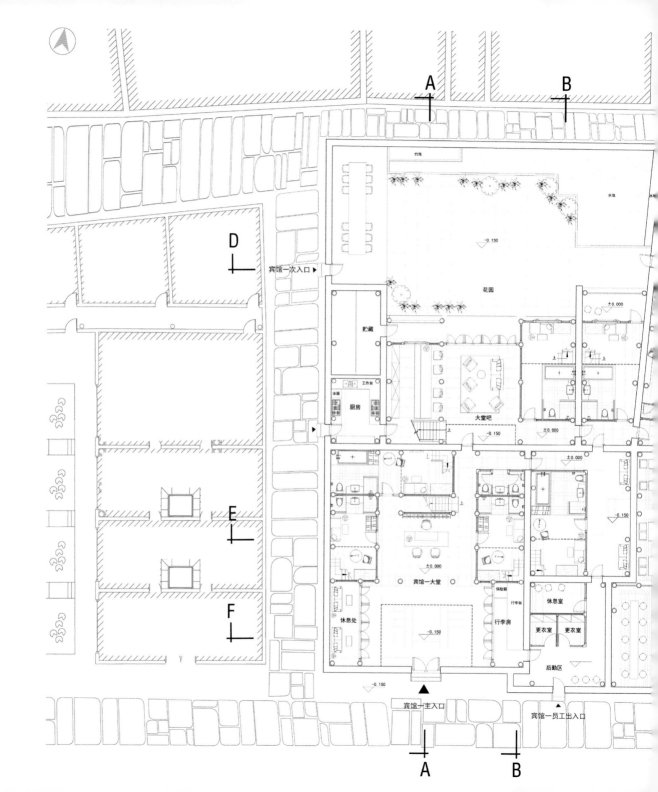

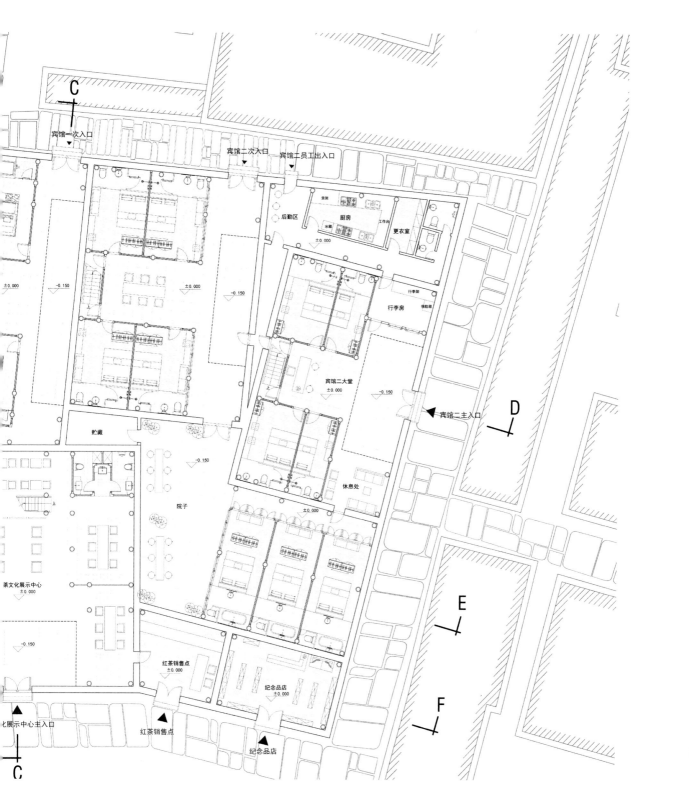

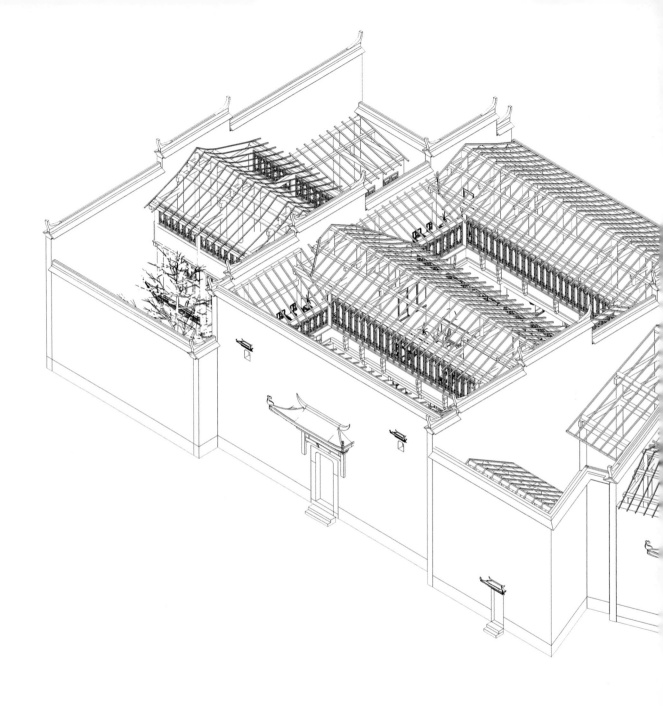

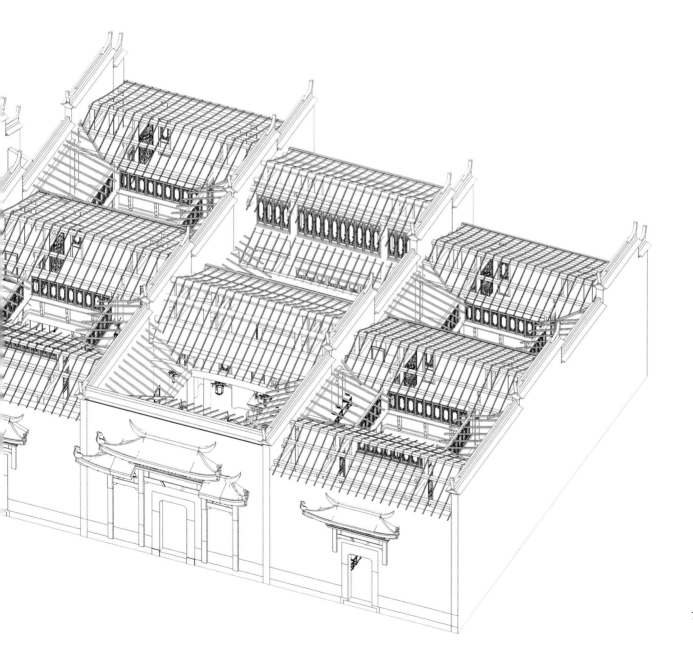

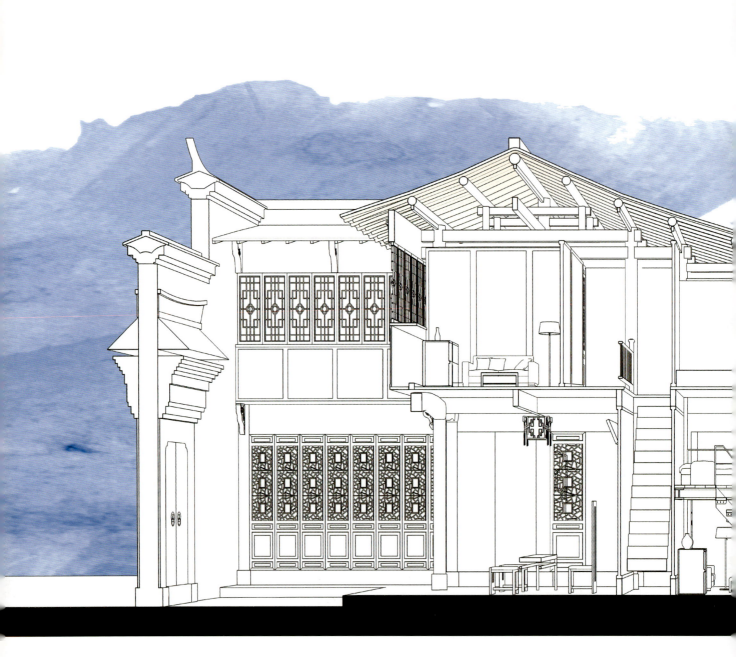

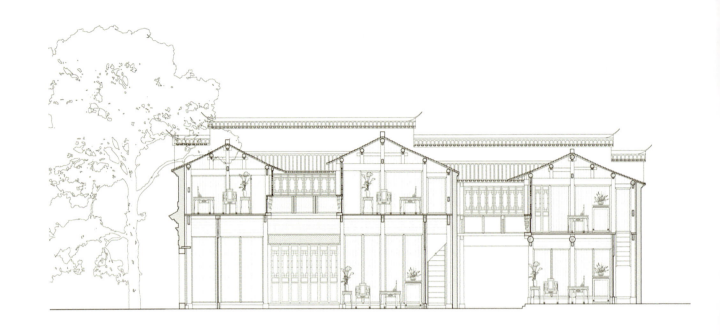
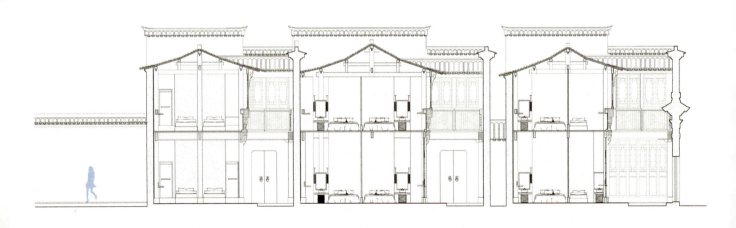

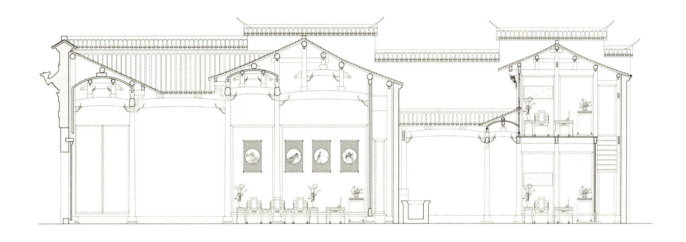
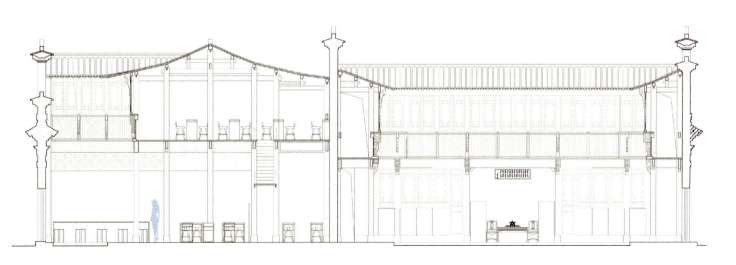

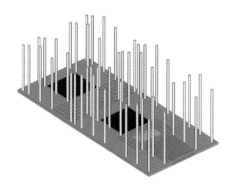 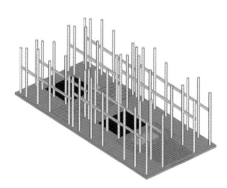 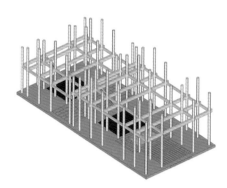
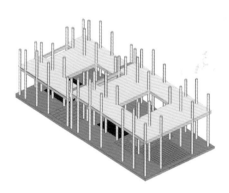 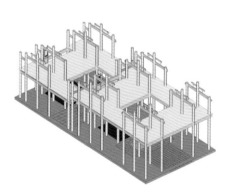 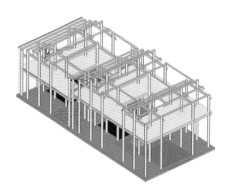
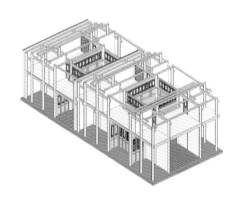 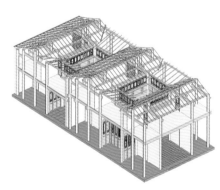 

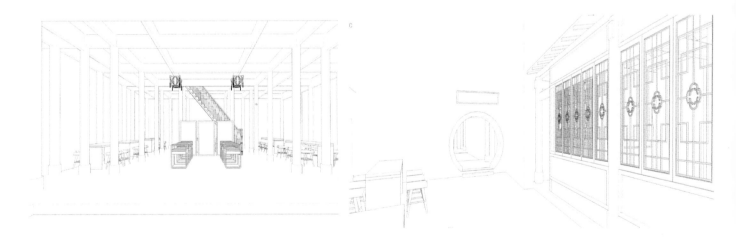

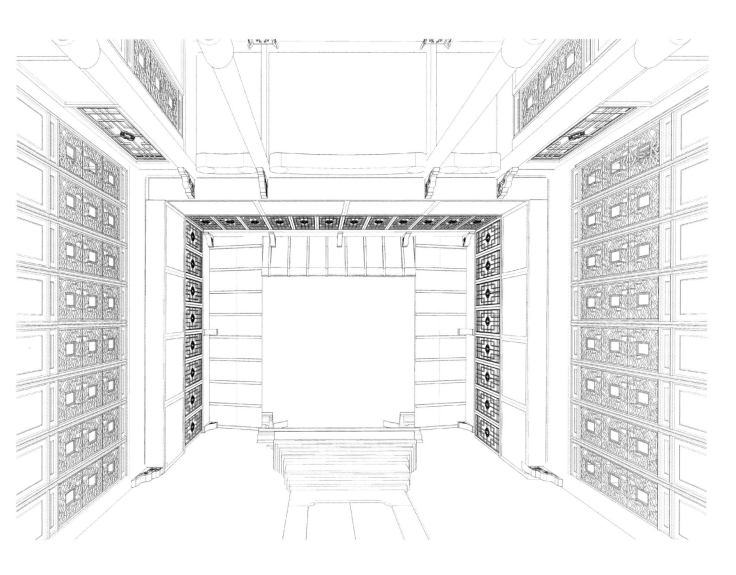

建筑设计研究（一） DESIGN STUDIO 1
# 南京老城南传统院落改建研究
RENOVATION OF THE NORTH SIDE OF SHENGZHOU ROAD AND BOTH SIDES OF PINGSHI LAND

张雷

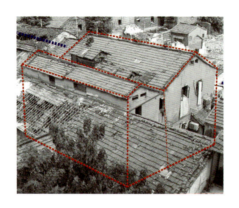

**教学目标**

　　课程从"环境"、"空间"、"场所"与"建造"等基本的建筑问题出发，通过南京老城南城市肌理和建筑类型的分析以及其后功能置换后使用空间的重新划分，从建筑与基地、空间与活动、材料与实施等关系入手，将问题的分析理解与专业的表达相结合，达到对建筑设计过程与设计方法的基本认识与理解。

**研究主题**

　　建筑类型、空间再划分、建筑更新、建造逻辑。

**设计内容**

　　对老城南升州路北侧评事街至大板巷段、评事街两侧进行调研，结合该区域改造规划，每组选择一个区域，每人选择一个院落，通过功能置换和整修改造，使其满足新的使用要求。

**Training Objective**

Based on the basic architectural problems such as environment, space and place, and construction, the course asks students to begin with analyzing the relationship of buildings and the site, space and behavior, materials and implementation, to understand the old city fabric and later displaced function, and combine the professional expression with the analysis of the architectural problems, so as to comprehend the basic architectural design process and design method.

**Research Subject**

Architectural types, space redefine, architectural renovation, tectonic logic.

**Design Content**

Based on the investigation of the site, each group chooses one court or region to make it satisfy the new requirement through the functional replacement and spatial re-ognization.

将民国的店屋改造为"影像记忆馆",尝试用现代建筑语言创造出具有流动性和展览性的空间。
Transform the shop houses in the Republic of China into the "video memory pavilion" and try to create mobility and exhibition spaces with the modern architectural language.

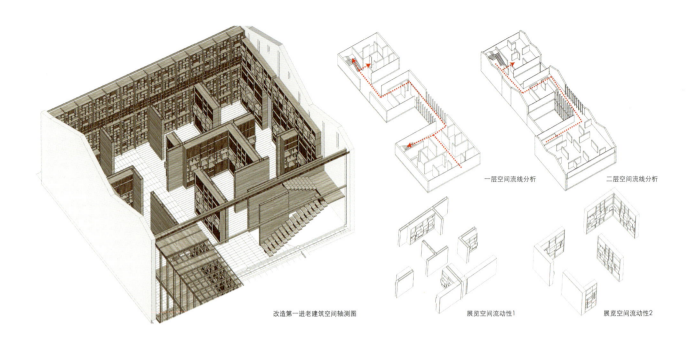

改造第一进老建筑空间轴测图　　一层空间流线分析　　二层空间流线分析　　展览空间流动性1　　展览空间流动性2

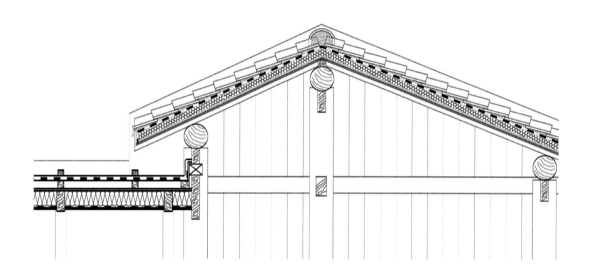

从传统建筑中提取出"格子"元素，对其进行几何抽象演变归纳出改造设计的基础语汇，建立新构件与原建筑的多重关系。

Extract the "grid" element from the traditional architecture, make geometric abstract evolution and summarize the basic vocabulary of transformation design.

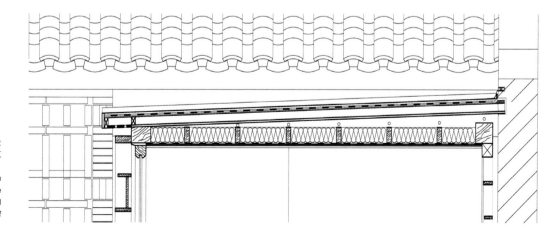

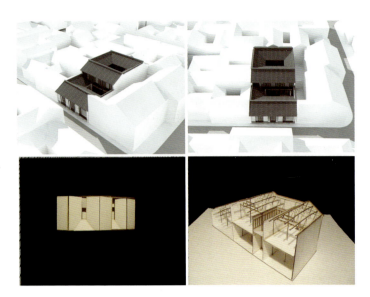

通过功能重组和空间重置，对传统院落空间的私密性和向心性加以调整，发掘其商业潜能。探索传统木构建筑与现代商业功能结合的可能性。
Adjust the privacy and centrality of traditional courtyard space through functional recombination and spatial resetting and explore their business potential. Explore the possibility of traditional wooden architecture combined with modern commercial function.

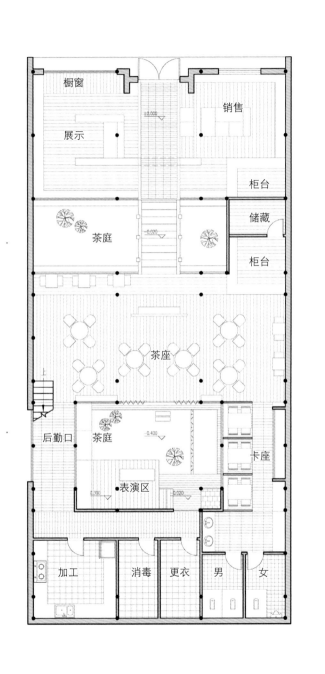
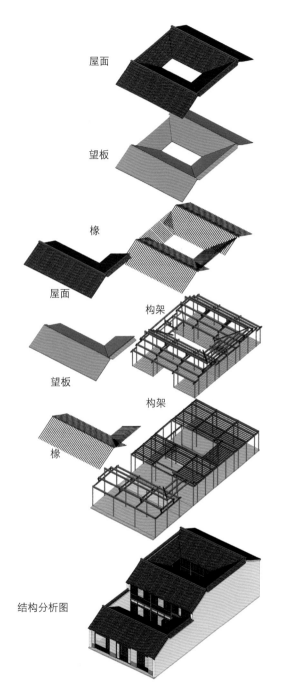

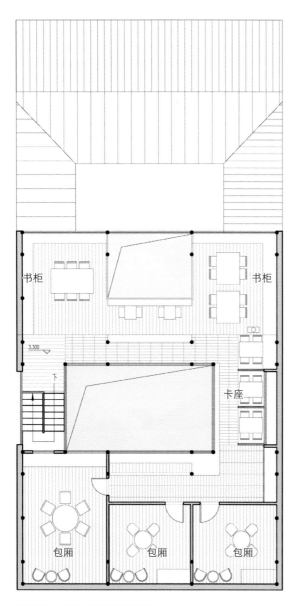

尝试将私密、内向的传统院落空间改造为公共、外向的商业空间。
Try to transform the private and introverted traditional courtyard space to public and extraverted commercial space.

建筑设计研究（二） DESIGN STUDIO 2

# 水乡聚落设计研究·呼吸作用
## WATER VILLAGE OF RESPIRATION

鲁安东

**背景**
基地：金坛市儒林镇汤墅村及下辖诸自然村
1. 位于长三角城市群内部缝隙地带的村庄，正在重新形成与城市的关系，因而面对"扩张"与"收缩"的双重不确定性。
2. 村庄的自然生长日益受到城市开发模式的干预，使得村庄原有的自维持和自调节能力衰竭。
3. 复杂的环境问题：极端气候、水灾、土地资源短缺、环境污染
4. 村落格局的变化：从滨水型向基础设施型

**概念**
空间—呼吸作用；社会—再兴；自然—抵抗力；建造—再利用。

**设计目标**
1. 基于"非正式"的村庄规划：利用基础设施支持呼吸作用，重新定位村庄与水的关系（水利与水害）；
2. 可持续的聚落设计：宅基地和自留地的调整、新农房设计、村庄加密；
3. 可持续的建筑设计：农房与公建——本地小机械的施工条件、可回收材料、被动式技术（厚墙）、绿色砌块和透水砖、生产性建筑（立体农业）。

**Background**
Site: Tangshu village, Rulin town, Jintan city
1.The village is located in a gap zone within the Yangtze Delta region. It is reformulating its relationship with nearby cities and therefore facing the two-fold uncertainty of expanding and shrinking.
2.The natural growth of the village is increasingly influenced by urban modes of development, which have exhausted its original capacity of self-maintenance and self-adjustment.
3.Complicated environmental issues: extreme climate, flood, scarcityof land resource, environmental pollution.
4.Change in the spatial form of the village: from water-front type to infrastructure-based type.

**Concept**
Space - respiration，society - revitalization，nature - resilience，construction - reuse.

**Design Objective**
1.A village plan based on "informality": use infrastructure to support respiration; reconfigure the relationship between the village and the water (the opportunities and threats of water).
2.Sustainable habitat design:
The adjustments to residential plots and reserved agricultural lands, design of new type of rural houses, densification of the village.
3.Sustainable architectural design:
Houses and public buildings: construction based on local machinery, recyclable materials, passive energy-saving techniques (e.g. thick wall), productive buildings (e.g. vertical agriculture).

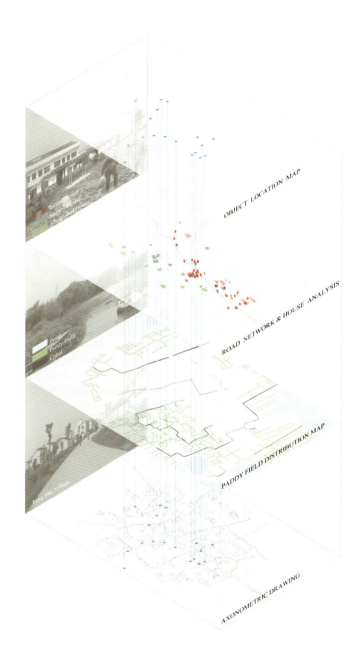

蜉蝣

通过对村庄中短暂存在的日常物品的空间微处理，以及利用个人移动设备进行叙事建构，本项目探索了基于物质空间、记忆和历史书写的新的村落更新模式。

Ephemera
By micro-intervention into the space of the short-lived everyday objects in the village, and by using personal mobile devices to fabricate their narratives, this project explores a new mode of village regeneration based on material space, memory and history-writing.

生态学
本项目从自然变迁的角度理解乡村聚落的当代更新。通过利用废弃建筑材料对小公共场所的针灸式改善，本项目为收缩过程的村落维持了空间的活力和品质。
Ecology
This project examines the regeneration of villages from the viewpoint of natural evolution. By reuse abandoned building materials to improve small public places in an acupunctural way, this project maintains the vitality and quality of the village in the process of shrinking.

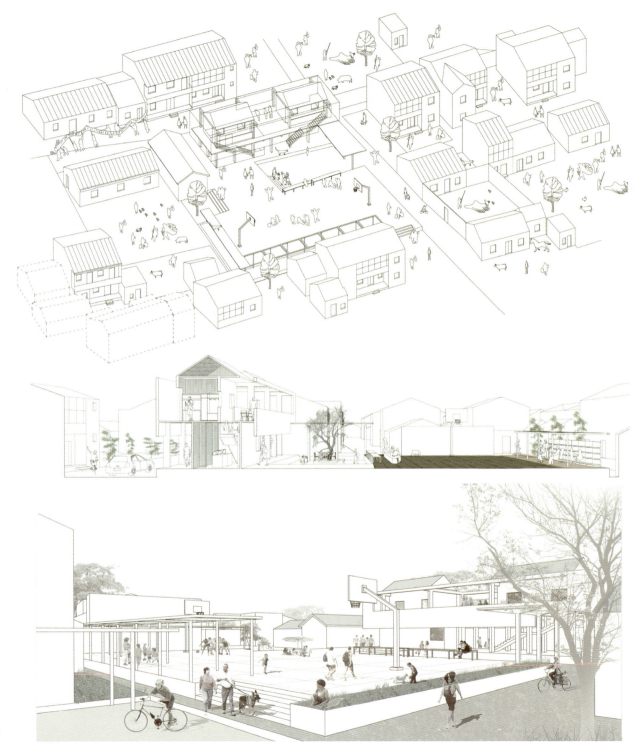

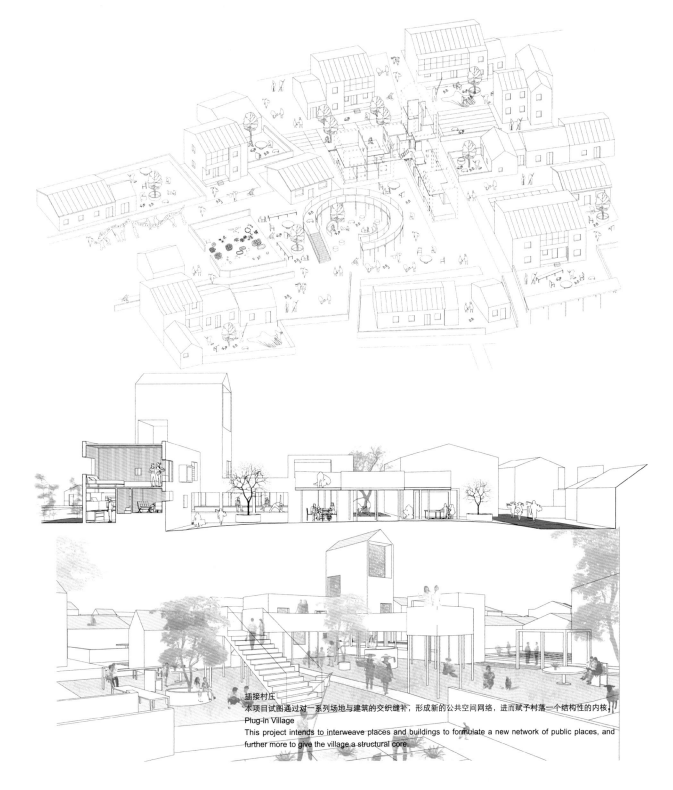

插接村庄
本项目试图通过对一系列场地与建筑的交织缝补，形成新的公共空间网络，进而赋予村落一个结构性的内核。
Plug-in Village
This project intends to interweave places and buildings to formulate a new network of public places, and further more to give the village a structural core.

建筑设计研究（二） DESIGN STUDIO 2
# 城市空间设计·层与界
FROM VOLUMN TO SPACE · MUTILLAYER & INTERFACE

丁沃沃

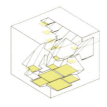

**教学目标**

　　城市化进程改变国土的自然地貌和人的生存空间。事实上，先行完成城市化的西方发达国家尤其是欧洲早已意识到平面扩张城市的恶果，转而在空间上寻求突破，由此引发了丰富多样的城市设计方法的探索。集约化城市的要义之一是空间的集约，然而高密度空间引发的问题往往制约了集约化城市的实现。本课题拟通过城市空间设计实验了解城市物质空间的本质和效能，初步掌握城市空间分配和城市建筑之间的关系。此外，通过城市空间设计练习进一步深化空间设计的技能和方法。

**研究主题**

　　以"层"与"界"作为操作元素，研究多维城市设计的方法，探索"多层化"的城市空间与城市地块指标体系之间的关系。

**设计内容**

　　1. 以南京新街口中心地块作为设计实验的场所，通过实地调研、案例研习、设计分析和试验，探讨高效城市空间的建构方法和内涵。

　　2. 观察与发现城市物质空间内的运作现象，基于建筑学图示方法，从空间意向出发建构城市物质空间的层级与界面，由"空间意向""扫描"出城市空间的"层"与"界"。

**Training Objective**

Chinese traditional villages and towns are fading away in direct proportion to Chinese urbanization speed, and the traditional farming culture, traditional handicraft and traditional values are on the verge of disappearing and weakening. The course focuses on studying the topic about Chinese traditional village and town protection and updating. Taking Hong Family's big house in Qimen, Anhui and surrounding land protection and updating plan for example, it makes basic research on traditional villages and towns and puts forward the concept planning of overall protection and updating on the research basis. Students understand the traditional villages and towns and traditional building culture through survey and planning design, learn planning knowledge and practice building design skills.

**Research Subject**

Study the method of multi-dimensional urban design with "layer" and "interface" as operation elements and explore the relationship between "multi-layer" urban space and urban plot indicator system.
Research topic
Context and Culture, Typology of Traditional Building, Master Planning for Preserving, Program and Function, Technique and Construction .

**Design Content**

1. Explore efficient urban space construction methods and connotation through field survey, case study, design analysis and experiment with Nanjing Xin Jiekou central plot as the design experiment place.
2. Observe and discover the operation phenomenon in the urban material space, start from the space intention and build thelayer and interface of urban material space on the basis of architectural graphic method, and "scan" the "layer" and "interface" of the urban space from "space intention".

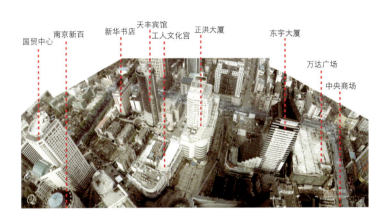

设计始于场地调研，其内容包括相关城市设计范围内及其周边的用地性质、交通状况、交通组织方式、建筑功能分布以及城市空间使用强度和城市活动环境评估。

Design begins with field survey and its contents include the land usage, traffic conditions, transportation organization modes and building function distribution as well as urban space use intensity and urban activity environment assessment in the related urban design scope and surrounding area.

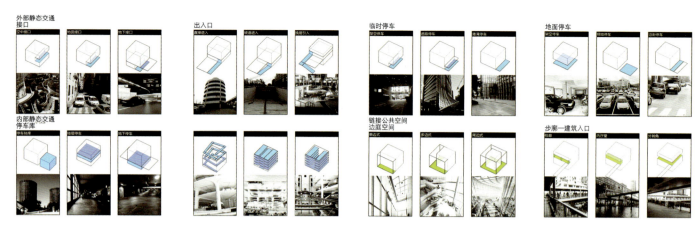

Field survey is spatial experience, while pertinent literature reading and summary can provide more high-density urban operation cases and operation examples. Initially understand the relationship between spatial types and urban activities and the allocation location of urban functions in urban space through literature reading and case analysis and put forward the spatial allocation ways of the case according to the survey content.

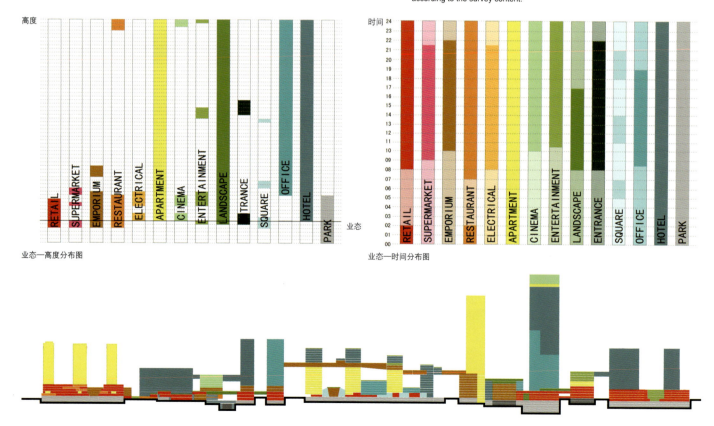

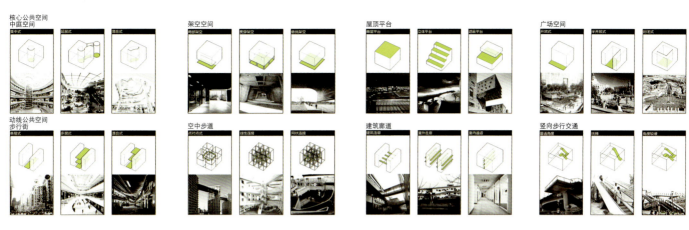

实地调研是空间的体验,而相关文献阅读和总结则能够提供更多的高密度城市运作的案例和操作范例。通过文献阅读和案例分析,初步了解空间类型和城市活动之间的关系、城市功能在城市空间中的分配位置,根据调研的内容提出本案的空间分配方式。

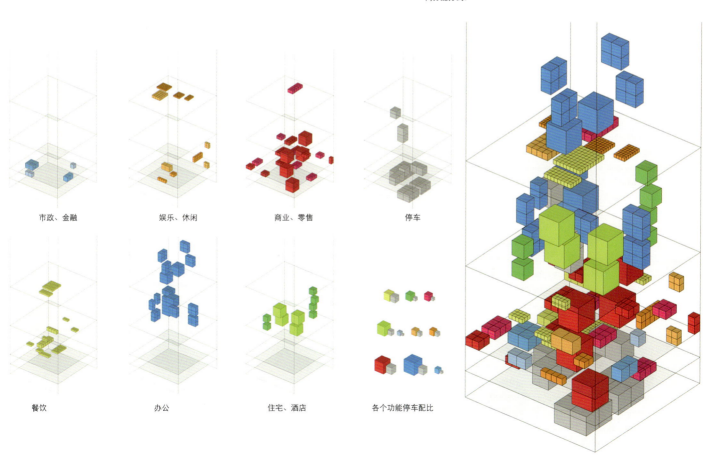

场地停车问题分析

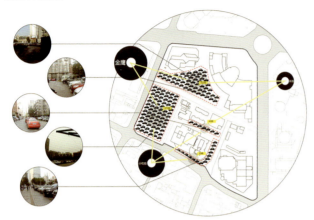

停车服务内外关系

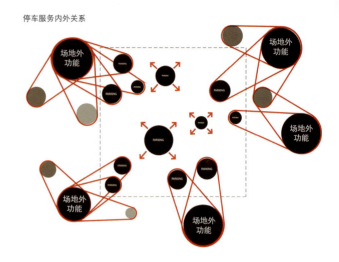

功能粘系剩余空间

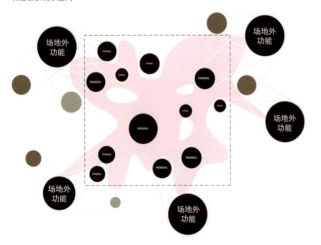

穿行人流意向分析

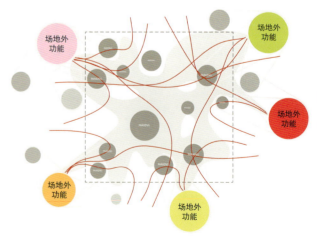

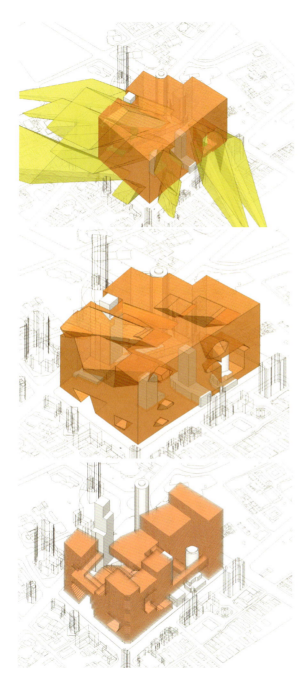
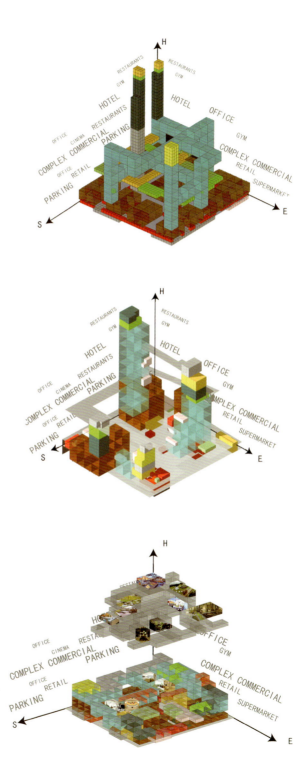

空间组织意向是城市设计的价值取向，多层交通组织方式、多层公共空间以及公共活动空间日照环境是本次设计主要考虑的因素。
The spatial organization intention is the value orientation of urban design. Multi-layer traffic organization ways, multi-layer public space and public activity space sunlight environment are the main factors considered in the design.

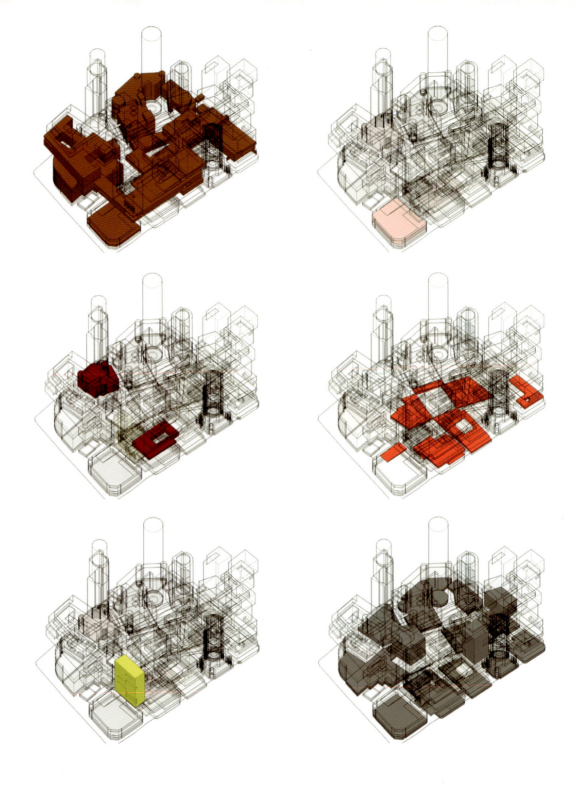

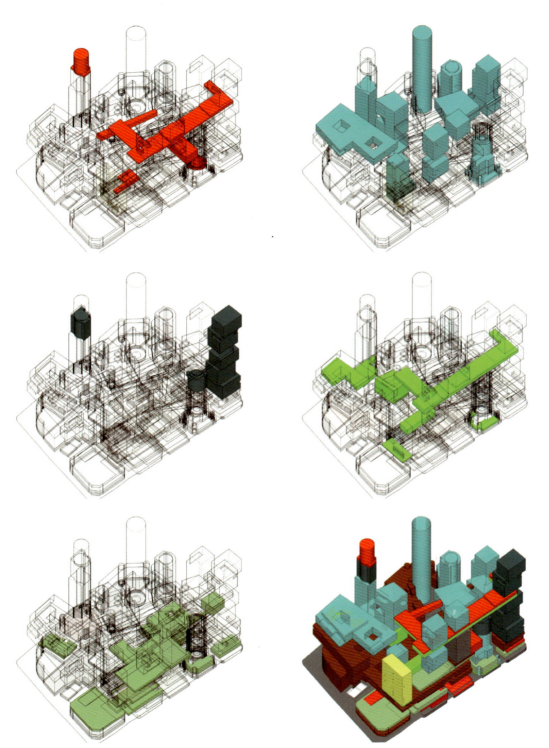

作为设计限定条件，城市设计场地中保留了现有的几座高层建筑，给设计带来了一定的挑战。新的功能设置和新的交通组织方式需要对现状作充分的分析，通过建筑的手段操作形体，构筑空间。

As the design limits, several high buildings are retained at the urban design site, which brings a certain challenges to design. New functional setting and traffic organization ways need to make sufficient analysis on the current situation and build spaces through the method operation form of buildings.

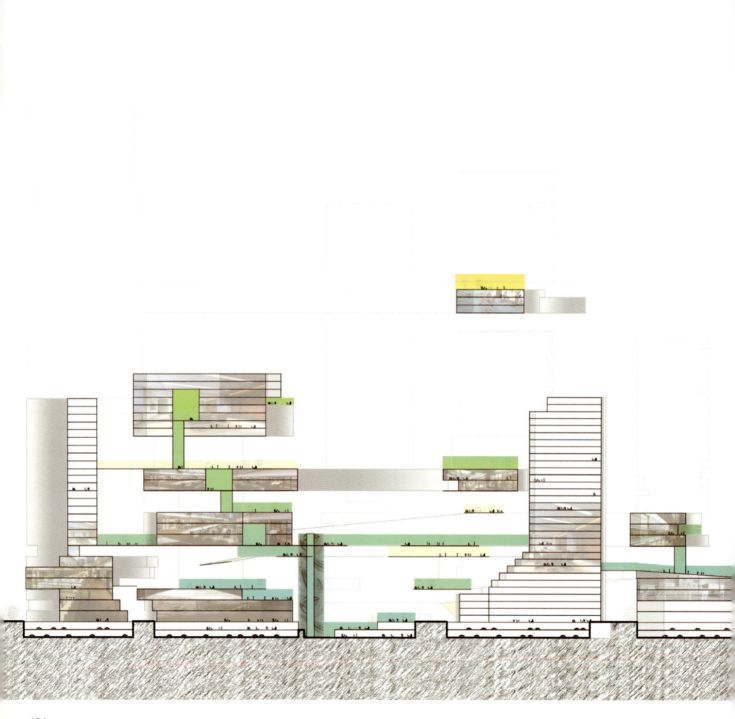

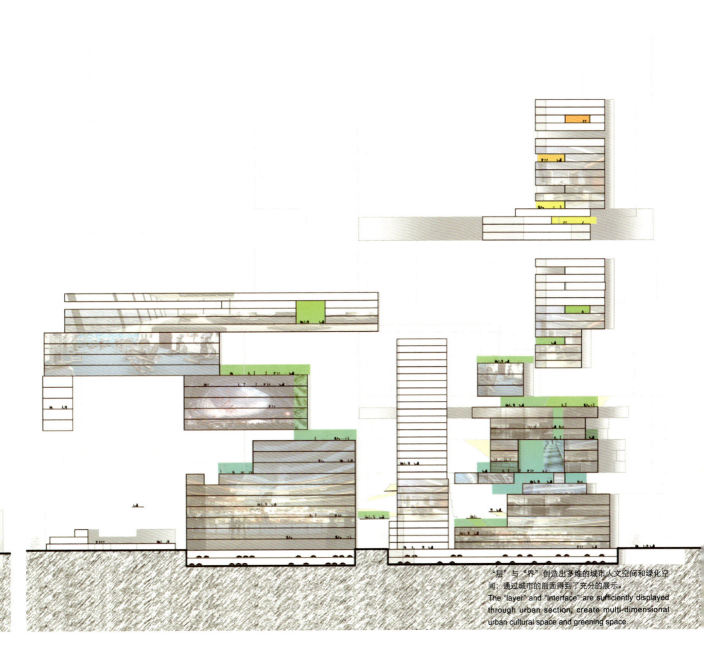

"层"与"界"创造出多维的城市人文空间和绿化空间；通过城市的剖面得到了充分的展示。
The "layer" and "interface" are sufficiently displayed through urban section, create multi-dimensional urban cultural space and greening space.

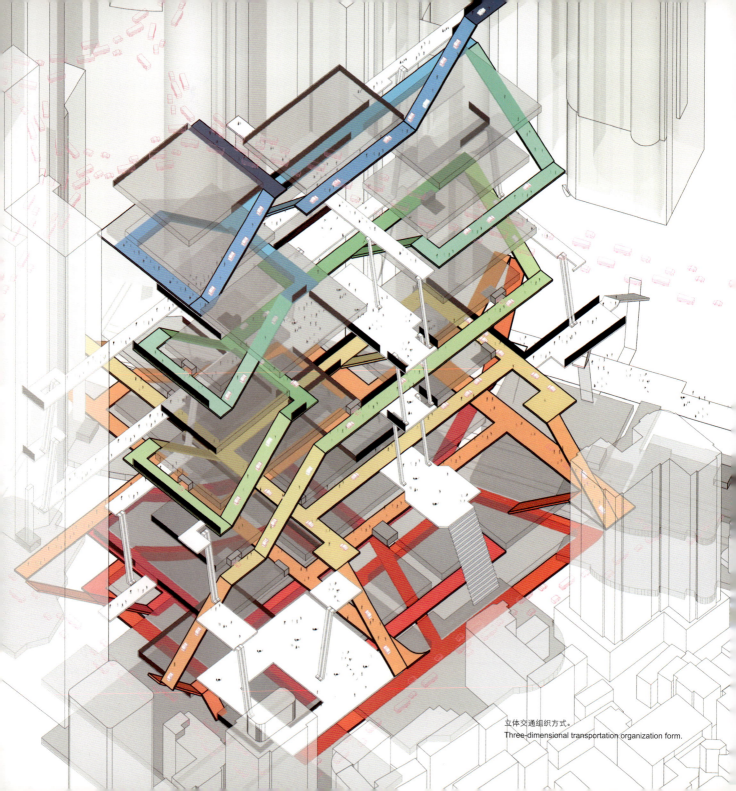

立体交通组织方式。
Three-dimensional transportation organization form.

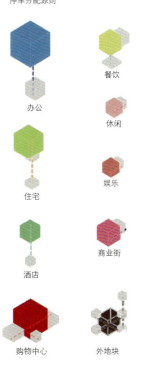

停车分配原则

办公 / 餐饮 / 住宅 / 休闲 / 酒店 / 娱乐 / 购物中心 / 商业街 / 外地块

多层公共空间组织方式。
Multi-layer public spatial organization form.

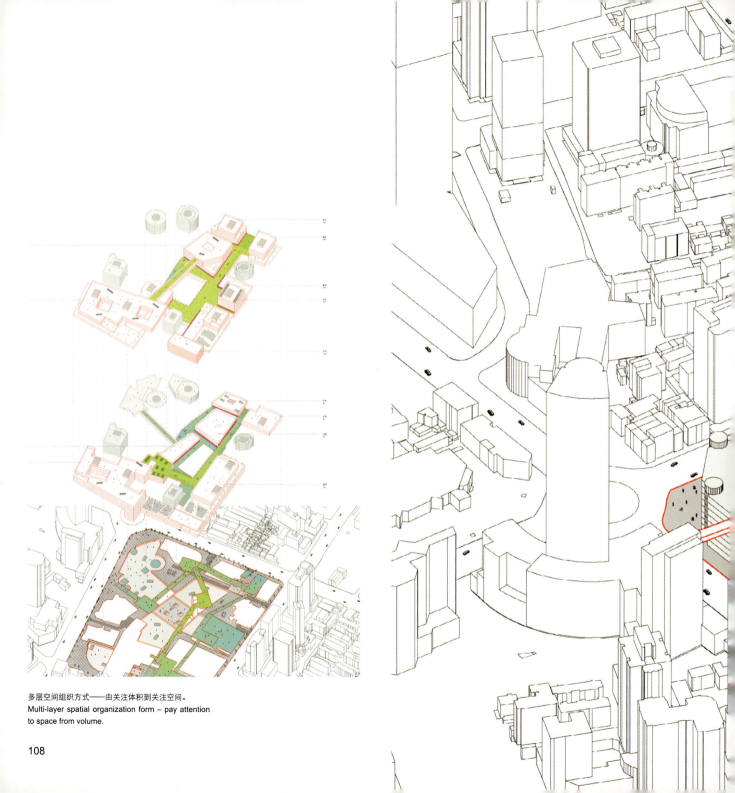

多层空间组织方式——由关注体积到关注空间。
Multi-layer spatial organization form – pay attention to space from volume.

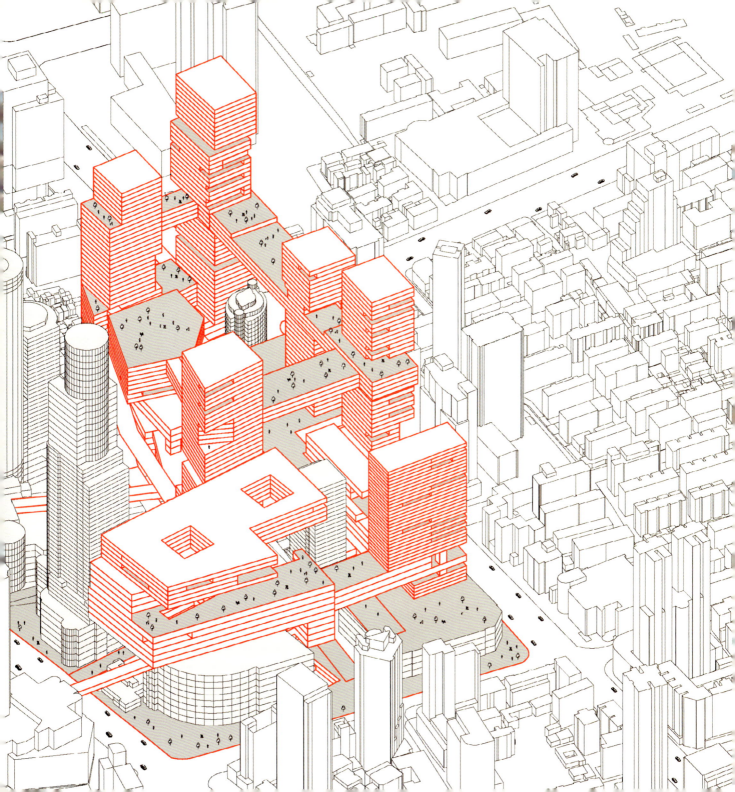

建筑设计研究（二） DESIGN STUDIO 2

# 基本建筑建构研究
## CONSTRUCTIONAL DESIGN

傅筱

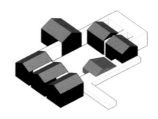

**教学目标**
设计概念与构造设计。

**设计内容与计划（2人/组）**
1. 设计概念与构造设计
（1）处理好节点的基本工程技术问题。
（2）根据设计概念研究建造材料的选用和节点设计，在满足基本功能性构造技术的前提下，重点研究超越功能性技术问题的构造设计表达。
时间：6周；成果：数字模型、概念分析图、构造详图。
2. 设计成果整理与表现
（1）1：100平面、立面图纸，表达深度达到施工图深度。
（2）剖面要求：先以节点大样的深度绘制1：20剖面图，然后再绘制表达
空间的单线剖面图1：100。抛弃施工图局部断面的表达方式。
（3）必须有带节点的空间透视表达。
（4）必须有带节点的三维轴侧表达。
时间：2周，成果：PPT演示和A1展板（不少于2张）。

**Training Objective**
Design concepts and structural design

**Design Content and Program**
1. Design concept and structural design (6 weeks)
(1) Settle the basic engineering technical issues regarding the nodes.
(2) Based upon the design concepts, study the selection of construction materials and the design of nodes; and on the premise of satisfying basic functional construction techniques, place emphasis on studying the expression of structural design beyond technical issues.
Achievements: digital models, conceptual analysis diagrams, detailed constructional drawings
2. Reorganization and expression of design results (2 weeks)
Requirements: The depth of expressions on large-scale plan drawings, elevation drawings and sectional drawings shall reach that of the construction drawings; the detailed drawing of nodes must have sectional nodes, plan nodes and 3-dimensional nodes; the expression must have spatial perspective expressions with nodes and 3-dimensional isometric expressions with nodes.
Achievements: PPT presentations and A1-format exhibition boards (at least 2 pieces).

通向二层平台的楼梯,通过楼梯材质和做法的精心处理，加强了上下空间的联系。
The upper and lower spatial relation is strengthened in the staircase leading to the second layer platform through elaborate processing of the material and practice of the staircase.

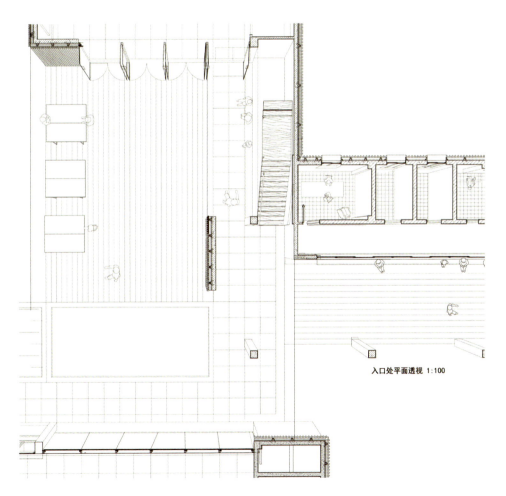

入口处平面透视 1:100

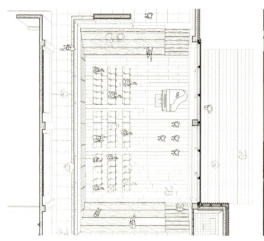

音体室空间：儿童身体接触处均为亲切的木质材料，顶棚暴露单向梁并结合自然光线，整体处理轻松自然。
Music and sports space: The places that children contact are all warm wood materials. The one-way beam exposed in the ceiling is united with the natural lighting, and the overall processing is natural and relaxing.

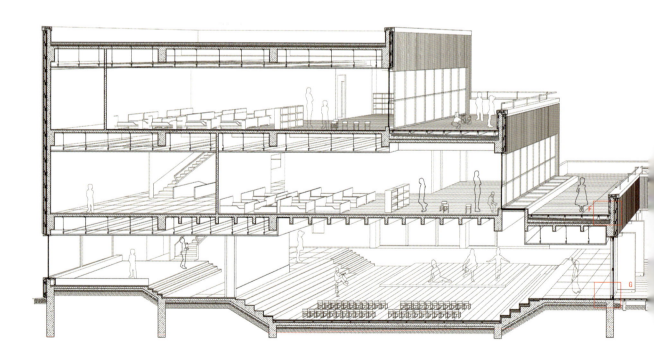

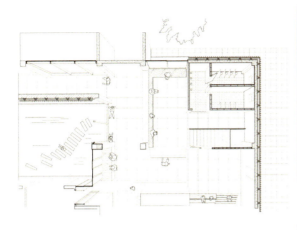

过渡空间
Transitional space

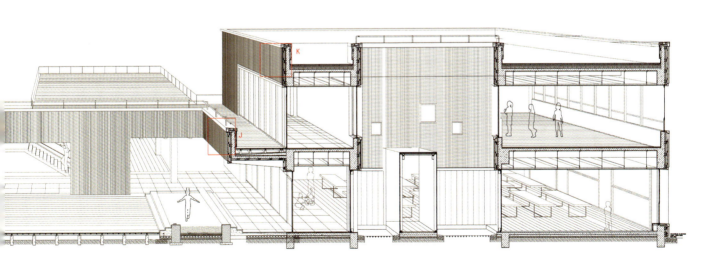

幼儿园空间剖透视：剖透视的训练可以让学生建立起空间、结构、节点、材质之间的关联。
The profile perspective of kindergarten: The training of profile perspective can make students set up the incidence relation among spaces, structures, nodes and materials.

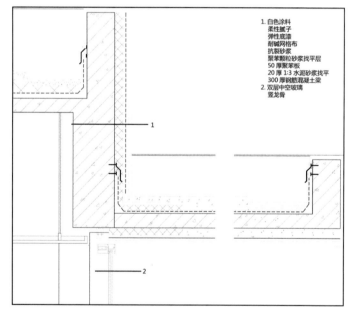

1. 白色涂料
   柔性腻子
   弹性底漆
   耐碱网格布
   抗裂砂浆
   聚苯颗粒砂浆找平层
   50 厚聚苯板
   20 厚 1:3 水泥砂浆找平
   300 厚钢筋混凝土梁
2. 双层中空玻璃
   竖龙骨

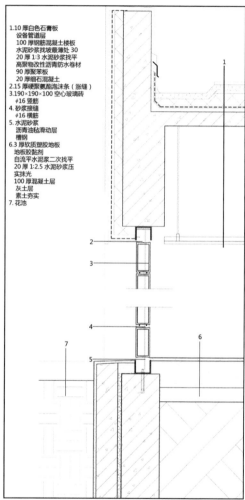

1. 10 厚白色石膏板
   设备管道层
   100 厚钢筋混凝土楼板
   水泥砂浆找坡最薄处 30
   20 厚 1:3 水泥砂浆找平
   高聚物改性沥青防水卷材
   90 厚聚苯板
   20 厚细石混凝土
2. 15 厚硬聚氨酯泡沫条（涨缝）
3. 190×190×100 空心玻璃砖
   φ16 竖筋
4. 砂浆接缝
   φ16 横筋
5. 水泥砂浆
   沥青油毡滑动层
   槽钢
6. 3 厚软质塑胶地板
   地板胶黏剂
   自流平水泥浆二次找平
   20 厚 1:2.5 水泥砂浆压
   实抹光
   100 厚混凝土层
   灰土层
   素土夯实
7. 花池

半透明墙体与花池植物关系大样。
The relationship bulk sample of the translucent wall and flower pond plants.

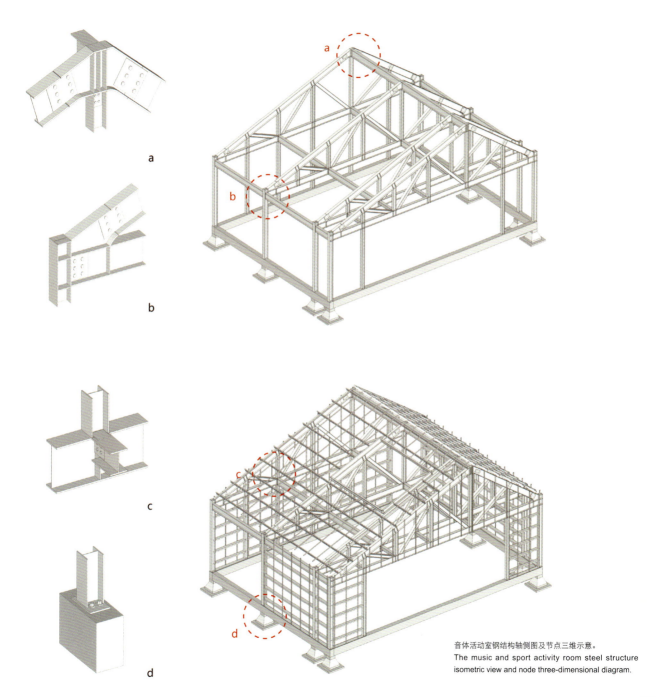

音体活动室钢结构轴侧图及节点三维示意。
The music and sport activity room steel structure isometric view and node three-dimensional diagram.

115

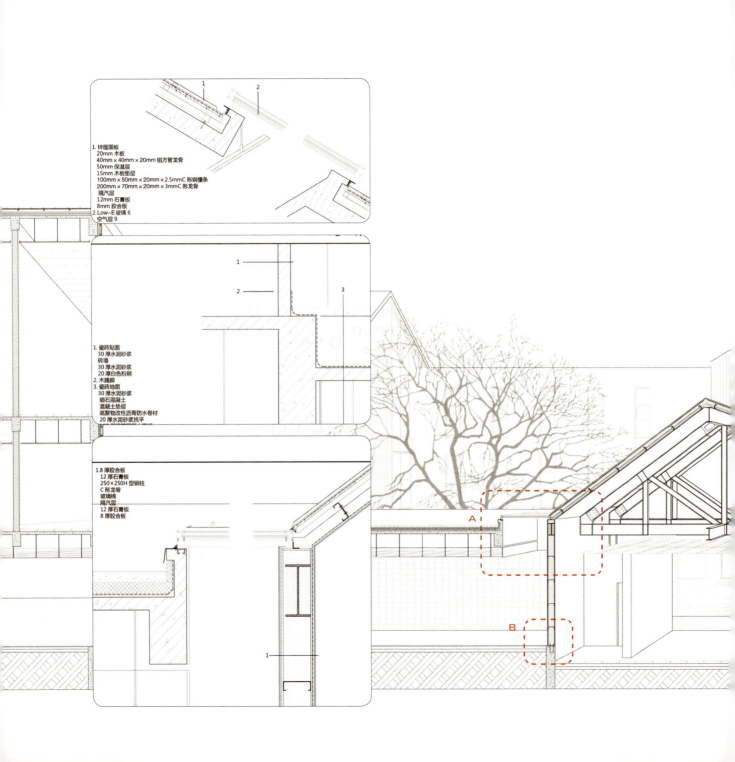

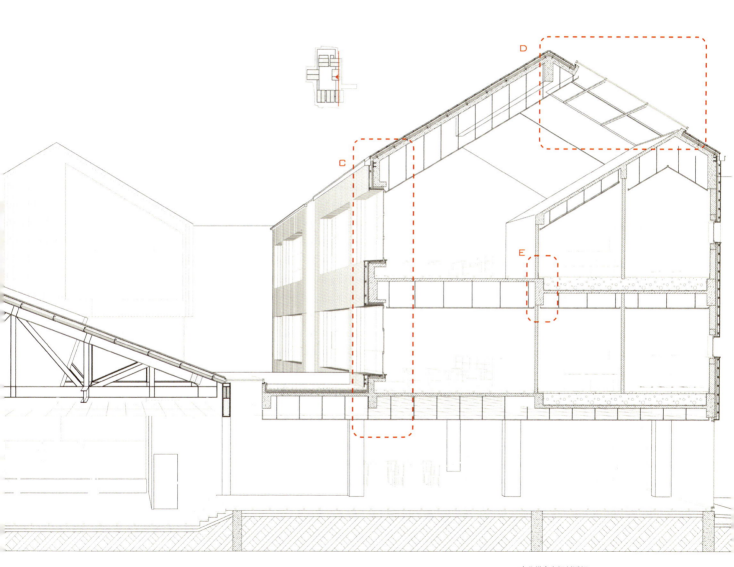

南北纵向空间剖透视
The north-south longitudinal space profile perspective

建筑设计课程
ARCHITECTURE DESIGN COURSES

本科二年级
## 建筑设计（一）小建筑设计
刘铨 冷天
课程类型：必修
学时／学分：64学时／4学分

Undergraduate Program 2nd Year
## ARCHITECTURAL DESIGN 1: SMALL PUBLIC BUILDING · LIU Quan, LENG Tian
Type: Required Course
Study Period and Credits: 64 hours/4 credits

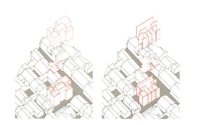

**课题内容**
老城古玩店设计

**教学目标**
根据建筑功能与外部空间的功能、流线需要，运用水平与垂直构件进行基本的空间限定及造型训练。

**研究主题**
学习利用水平构件与垂直构件形成并组织室内外空间；
学习出入口设置与简单的流线组织；掌握基本的人体活动与空间、构件的尺度关系；注意室内空间与相邻建筑、街道或庭院之间的空间及界面关系。

**设计内容**
在城市历史街区中设计具有商业与展示功能的小型建筑单体。在地块内现存建筑物已被拆除的假定前提下，独立完成一次完整的小型建筑单体方案的设计任务。建筑功能为具有展示性质的古玩商店，包括展示与销售、顾客接待空间以及值班室、洗手间、垂直交通等辅助功能空间，建筑层数二层，建筑面积160—200㎡，不得再开挖地下室获取额外的使用空间；楼梯、门窗等建筑构件的设计需满足基本建筑规范的要求。

**Subject Content**
Design of an Antique Shop in the Historic Urban District
**Training Objective**
To conduct space limit and modeling exercise with the horizontal and vertical components based on the functions of the building and external space as well as the streamline requirements.
**Research Subject**
Learn to model and organize the indoor and outdoor spaces with the horizontal and vertical components; study entrances and exits settings and simple streamline organization; understand the scale relation between the basic human activities and the spaces as well as components; and pay attention to the space and interface relations between the indoor space and the adjacent buildings, streets or courtyards.
**Design Content**
Design a small building with commercial and display functions in an urban historic block. Given the precondition that the existing buildings on the plot have been demolished, independently complete a small building design plan. The building is a two-story antique shop with the display function and a building area of 160-200m², containing spaces for display, sales and customer reception as well as the auxiliary functional spaces such as the duty room, rest room and vertical communication facilities. Extra using space cannot be acquired by digging a basement; design of the building components such as stairs, doors and windows should meet the requirements of the basic building codes.

---

本科二年级
## 建筑设计（二）小建筑设计
刘铨 冷天
课程类型：必修
学时／学分：64学时／4学分

Undergraduate Program 2nd Year
## ARCHITECTURAL DESIGN 2: SMALL PUBLIC BUILDING · LIU Quan, LENG Tian
Type: Required Course
Study Period and Credits: 64 hours/4 credits

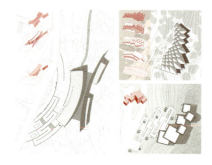

**课题内容**
风景区茶室设计

**教学目标**
训练理解物质环境与建筑空间生成的关系。

**设计内容**
基地位于紫金山风景区内一处临水的丘陵坡地，同时建筑功能也较为简单。坡度、坡向、视觉景观条件是建筑空间生成的主要因素。
1. 在给定的两块用地中选定一块进行设计，在地形条件解读的基础上，形成场地环境布局和建筑形体；
2. 建筑功能为风景区茶室，包括可供至少60人使用的大厅和30人使用的雅座，以及操作间、休息室、洗手间等必要的辅助用房，总建筑面积不超过300㎡，建筑应考虑无障碍设计；
3. 设计中应考虑两个车位的停车空间，如设计车库，不计入建筑面积；
4. 在综合考虑建筑内外功能与流线的情况下，对场地景观环境进行再组织；
5. 综合运用SketchUp、纸质模型、草图等工具分析理解地形以辅助设计。

**Subject Content**
Design of a Tea Shop in the Scenic Spot
**Training Objective**
To understand the relationship between the physical environment and architectural space generation.
**Design Content**
The base is a waterside hilly slope in the Purple Mountain Scenic Spot, and the building function is simple. The gradient, aspect of the slope and visual landscape conditions are the main factor of architectural space generation.
1. Select a land for design from the two designated lands. Generate the site environment layout and architectural form upon interpretation of the terrain conditions;
2. The building is tea shop in the scenic spot with a building area up to 300m², containing a hall for at least 60 people, private rooms for at least 30 people as well as the necessary auxiliary rooms such as the operation room, lounge and rest room. Accessible design should be considered;
3. The parking space with two stalls should be considered. The garage should not be included in the building area;
4. Reorganize the landscape environment of the site upon consideration of the external and internal functions and streamline of the building;
5. Analyze and understand the terrain with tools like SketchUp, paper models and sketches for design assist.

本科三年级
建筑设计研究（三）小建筑设计
周凌 童滋雨 钟华颖
课程类型：必修
学时/学分：72学时/4学分

Undergraduate Program 3rd Year
ARCHITECTURAL DESIGN 3: SMALL PUBLIC BUILDING · ZHOU Ling, TONG Ziyu, ZHONG Huaying
Type: Required Course
Study Period and Credits: 72 hours/4 credits

课题内容
赛珍珠纪念馆扩建
教学目标
此课程训练最基本的建造问题，使学生在学习设计的初始阶段就知道房子如何造起来，深入认识形成建筑的基本条件：结构、材料、构造原理及其应用方法，同时课程也面对场地、环境和功能问题。训练核心是结构、材料、场地。在学习组织功能与场地同时，强化认识建筑结构、建筑构件、建筑围护等实体要素。
文脉：充分考虑校园环境、历史建筑、校园围墙以及现有绿化，需与环境取得良好关系。
退让：建筑基底与投影不可超出红线范围。若与主体或相邻建筑连接，需满足防火规范。
边界：建筑与环境之间的界面协调，各户之间界面协调。基底分隔物（围墙或绿化等）不超出用地红线。
户外空间：扩建部分保持一定的户外空间，户外空间可在地下。
地下空间：充分利用地下空间。

设计内容
基地内地面最大可建面积约100 m²，地下可建面积200—300 m²，总建筑面积约300—400 m²，建筑地上1层，限高4m，地下层数层高不限。展示区域200—300 m²，导游处10 m²，纪念品部30 m²，茶餐厅60 m²，厨房区域 >10 m²，门厅与交通，卫生间。

Subject Content
Expansion of Pearl Buck's House
Training Objective
This course trains the students to solve the basic construction of architecture. Students should learn how to build an architecture at the very beginning of their studying, understand the basic aspects of architectures: the principles and applications of structure, material and construction. The course also includes the problem of site, enviroment and function. The keypoints of the course include site, structure and material. Students should strengthen the understanding of physical elements including structures, components and façades while learning to organize the function and site.
Context: The enviroment, historical building, the edge of the campus and the green belt around the site should be taken into consideration. The expansion is expected to have a good relationship with the surroundings.
Retreat Distance: The new architecture can't beyond the red line. Fire protection rule should be complied.
Boundary: Both the boundary between different buildings and between building and environment should be harmonized.
Open Space: Open space should be considered, which permitted to be placed underground.
Underground Space: Underground space should be well used.
Design Content
The maximum ground can be used in the base area is about 100m², while underground construction area is about 200-300 m², and the total floor area of architecture should be about 300-400m². The architecture should be 1floor above the groud lower than 4m.The underground levels have no limitation.Exhibition area: 200-300m²,Information center: 10m²,Shop: 30m²,Coffee bar: 60m²,Kitchen: >10m²,Lobby & walking space,Toilet .

本科三年级
建筑设计研究（四）中型公共建筑设计
周凌 童滋雨 钟华颖
课程类型：必修
学时/学分：72学时/4学分

Undergraduate Program 3rd Year
ARCHITECTURAL DESIGN 4: PUBLIC BUILDING · ZHOU Ling, TONG Ziyu, ZHONG Huaying
Type: Required Course
Study Period and Credits: 72 hours/4 credits

课题内容
大学生活动中心设计
教学目标
课程主题是"空间"，学习建筑空间组织的技巧和方法，训练空间的效果与表达。空间问题是建筑学的基本问题，课题基于复杂空间组织的训练和学习。从空间秩序入手，安排大空间与小空间、独立空间与重复空间，区分公共与私密空间、服务与被服务空间、开放与封闭空间。训练重点是空间组织，包括空间的秩序、空间的内与外、空间的质感及其构成等。以模型为手段，辅助推敲。设计分阶段体积、空间、结构、围合等，最终形成一个完整的设计。

设计内容
1. 空间组织原则
空间组织要有明确特征，有明确意图，概念要清楚，并且满足功能合理、环境协调、流线便捷的要求。注意三种空间：聚散空间（门厅、出入口、走廊）；序列空间（单元空间）；贯通空间（平面和剖面上均需要贯通，内外贯通、左右前后贯通、上下贯通）。
2. 空间类型
多功能空间、展示空间、专属空间、休闲空间、服务空间、交通空间。
总建筑面积控制在3000 m²以内，层数控制在4层以内。

Subject Content
Design of the College Student Center
Training Objective
The theme of the course is space. It trains students' skills of space organization and make them master methods of space organization and presentation. Space issues are the basic issues of architecture. This course organizes trainings and studies based on complex space. Students start with spatial order, arrange the large spaces and small spaces, independent spaces and repetitive spaces, differentiate public and private spaces, service and serviced spaces, open and close spaces. The key point of training is space organization, including the order of space, inner and outer, texture of space and its composition, etc. Models should be used as means to assisting deliberation. The course includes stages of volume, space, structure, enclosing, then forms a complete design.
Design Content
1. Principle of space organization
The space organization should have clear characteristics, intentions and concepts, while satisfying the requirements for reasonability, environmental harmony and convenient circulation. Gathering and dispersing spaces (hallway, entrance/exit, corridor); Sequence space (unit spaces); Through space (on plans and sections, including internal/external through, left / right / front / rear through, upper/lower through).
2. Type of space
Multi-function space,exhibition space,proprietary space,recreation space,Service space,circulation space: lobby, corridors, etc.
The total floor area ≤ 3,000m², ≤ 4 floors.

本科三年级
**建筑设计（五＋六）大型公共建筑设计**
华晓宁　王丹丹　钟华颖
课程类型：必修
学时 / 学分：144 学时 / 8 学分

Undergraduate Program 3rd Year
**ARCHITECTURAL DESIGN 5-6: COMPLEX BUILDING** · HUA Xiaoning, WANG Dandan, ZHONG Huaying
Type: Required Course
Study Period and Credits: 144 hours/8 credits

**课题内容**
　　城市建筑——社区商业中心＋观演中心设计
**教学目标**
　　训练学生分析和理解较为复杂的城市关系，并在设计中做出应对。初步掌握较为复杂的大型公共建筑的功能、空间与流线组织。
**研究主题**
　　实与空，内与外，层与流，轴与界。
**设计内容**
　　在用地上布置社区商业中心（约 30000 m²）、观演中心（不超过 10000 m²）和至少一个不小于 6500 m² 的城市广场，并生成其他相应的城市居民行动路径和外部公共空间。

**Subject Content**
Urban Buildings-Design of Community Business Center and Performance Center
**Training Objective**
To analyze and understand the complex urban relationships and react to them in design, and to preliminarily understand the functions, spaces and streamline organizations of the complex large public buildings.
**Research Subject**
Solid and empty, internal and external, layer and flow, as well as axis and border.
**Design Content**
Design a community business center (about 30000m²), a theater (up to 10000m²) and at least one city square (at least 6500m²) on the land, and generate corresponding paths and external public spaces.

---

本科四年级
**建筑设计（七）高层建筑设计**
吉国华　胡友培　尹航
课程类型：必修
学时 / 学分：72 学时 / 4 学分

Undergraduate Program 4th Year
**ARCHITECTURAL DESIGN 7: HIGH-RISING BUILDING** · JI Guohua, HU Youpei, YIN Hang
Type: Required Course
Study Period and Credits: 72 hours/4 credits

**课题内容**
　　高层办公楼设计
**教学目标**
　　高层办公建筑设计涉及城市、空间、形体、结构、设备、材料、消防等方面内容，是一项较复杂与综合的任务。本课题采取贴近真实实践的视角，教学重点与目标是帮助学生理解、消化以上涉及各方面知识，提高综合运用并创造性解决问题的技能。
**设计内容**
　　建筑容积率≤7，建筑限高≤100 m，裙房高度≤24m，建筑密度≤50%；基底南侧、东侧为规划道路，在假设其存在基础上，展开方案设计。
　　高层部分为办公楼，设计应兼顾各种办公空间形式。裙房设置会议中心，须设置 400 人报告厅一个，100 人报告厅 4 个，其他各种会议形式的中小型会议室若干，以及休息厅、服务用房等。会议中心应独立对外使用。地下部分主要为车库和设备用房。

**Subject Content**
Design of High-rise Office Building
**Training Objective**
Design of the high-rise office building is a complicated and comprehensive task, involving city, space, form, structure, equipment, materials and fire control. From a perspective close to the practice, this course focuses on and aims at helping students understand and grasp the knowledge of the above-mentioned aspects and improving their skills of integrated use and creatively solving problems.
**Design Content**
Building floor area ratio ≤ 7, building height limit ≤ 100m, podium height ≤ 24m, building density ≤ 50%; the south and east sides of the base are planned road, and the design can be based on their existence.
The high-rise building is an office building, so the design should consider various forms of the office space.
The conference center should be set in the annex, with a 400-seat lecture hall, four 100-seat lecture halls, several medium- and small-sized meeting rooms of various forms, lounges, and service rooms. The conference center should be available for independent external use. The basement should mainly consist of the garage and equipment room.

本科四年级
**建筑设计（八）城市设计**
胡友培  尹航
课程类型：必修
学时 / 学分：72 学时 / 4 学分

Undergraduate Program 4th Year
**ARCHITECTURAL DESIGN 8: URBAN DESIGN** · HU Youpei, YIN Hang
Type: Required Course
Study Period and Credits: 72 hours/4 credits

课题内容
金城厂街区城市设计——对大街区城市形式的研究设计
教学目标
1. 掌握城市设计的基本技能。
2. 培养对城市形态、肌理形态、建筑类型之间内在关系的认知，掌握通过类型学设计方法，操作城市肌理、城市形态的技能。
3. 重点训练对地块建设指标与城市形态间内在的复杂联系之认知，并初步建立起一种量、形结合的城市形态科学认识论与设计方法论。
4. 培养创造性的、具有想象力的解决城市现实问题的气质与能力。
研究主题
1. 引入城市肌理作为城市设计工具，在规划指标的"量"与建筑的"形"之间建立更有机的链接。
2. 探索大街区更新改造的可能城市形式，在多样性与类型化（秩序统一）之间建立平衡。
关键词：城市肌理、类型、功能混合。
设计内容
1. 至少提供南北、东西向城市支路各一条，以缓解大街区周边城市主干路网的交通压力。
2. 控制总容积率及功能构成与上位控规一致。
3. 每个地块内至少具有两种以上功能。

**Subject Content**
Urban Design of the Block of Jincheng Factory—Study on Design of the Urban Form of the Dajie Area
**Training Objective**
1. To master the basic skills of urban design.
2. To recognize the intrinsic relationship among the urban form, fabric pattern and building type, and to acquire the skill of designing the urban fabric and form with the typological design method.
3. To recognize the complicated intrinsic relationship between the block construction indexes and urban form, and to preliminarily establish a scientific epistemology and design methodology of urban form integrating quantity and form.
4. To develop the temperament and capability of creatively and imaginatively solving the practical problems of the cities.
**Research Object**
1. Introduce the urban fabric as an urban design tool to establish a more organic connection between the "quantity" of the planning index and the "form" of the buildings.
2. Explore the possible urban forms for Dajie area renewal and balance diversity and typification (order unity).
**Design Content**
1. Provide at least one urban branch in south-north and east-west directions each, so that the traffic pressure of the urban trunk road network around Dajie area can be relieved.
2. Total volume rate control and functional form are consistent with the regulatory planning.
3. Each plot has at least two functions.

---

本科四年级
**毕业设计**
赵辰  窦平平
课程类型：必修
学时 / 学分：1 学期

Undergraduate Program 4th Year
**THESIS PROJECT** · ZHAO Chen, DOU Pingping
Type: Obligatory
Study Period and Credits: 1 term

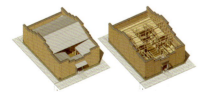

课题内容
福建山区传统村落保护与复兴项目规划设计
教学目标
掌握建筑设计基本的技能与知识（测绘、建模、调研、分析），并能对特定的地域和历史建筑进行深入的设计研究（内容策划，建筑结构、构造），根据社会发展的需求，提出改造和创造的可能。
设计内容
本次毕业设计针对屏南县北村、政和县杨源、锦屏等村落进行规划设计。在选定的村落现状研究的基础上，进行村落景观空间的整体规划。并且，选择相关重点区域与建筑，进行专项的建筑设计。

**Subject Content**
Project for the Preservation and Rehabilitation of Traditional Villages in the Mountainous Areas of Fujian Province
**Training Objective**
To acquire basic skills and knowledge (mapping, modeling, research and analysis) about architectural design, be able to further design and study (content planning, building structure, and construction) region-specific and history-specific architecture, and put forward ideas about renovation and creation based on the demands of social development.
**Design Content**
This Thesis Project is to plan and design for North Village, Pingnan County as well as the villages of Zhenghe County (such as Yangyuan Village and Jinping Village). We will make overall plans for the landscape spaces of the selected villages based on the research into their status quo and give special architectural designs to the selected key areas and buildings.

本科四年级
**毕业设计：建筑技术科学专门化设计**
华晓宁　郜智
课程类型：必修
学时/学分：1学期

## Undergraduate Program 4th Year
### THESIS PROJECT: SPECIALIZED DESIGN OF BUILDING SCIENCE · HUA Xiaoning, GAO Zhi
Type: Obligatory
Study Period and Credits: 1 term

课题内容
南京河西城市生态公园生态展示馆及未来生态家建筑设计
教学目标
掌握整合于建筑设计策略中的绿色建筑相关设计理念、知识和方法。
设计内容
为普及低碳生态智慧理念，推动低碳生态智慧技术应用，创建国家级的低碳生态智慧样本城，南京决定建设河西城市生态公园。城市生态公园定位为低碳生态智慧技术的集中展示场所和国家级的低碳生态智慧科普教育基地。生态展示馆和未来生态家建筑组群位于生态公园内，希望将建筑、展陈和场地三者统一协调，均用作为低碳生态智慧理念和技术的集中展示场所。
本课题将开展以建筑绿色节能技术为主题的建筑设计与专题研究，成果以项目设计及研究报告的形式呈现。

### Subject Content
Architectural Design for the Ecological Exhibition Hall and Future Eco-home in Nanjing Hexi Urban Ecological Park
### Training Objective
To acquire green building-related design concepts, knowledge and methods integrated into architectural design strategy.
### Design Content
Nanjing decided to build up the Hexi Urban Ecological Park to popularize the low-carbon, ecological and smart concepts, driving the application of low-carbon, ecological and smart technologies and setting up a national example for a low-carbon, ecological and smart city. The urban ecological park is to be an exhibition place for the low-carbon, ecological and smart technologies and a national base for the popularization and education of these technologies. The building group of the Ecological Exhibition Hall and Future Eco-home are located in the park. The buildings, exhibition design and spaces are expected to coordinate with each other as a centralized exhibition place for the low-carbon, ecological and smart concepts and technologies.
This subject is to conduct an architectural design and case study with the theme of green and energy-saving architectural technologies, and the result will be displayed as the project design and study report.

本科四年级
**毕业设计：数字化建造技术专门研究**
童滋雨　钟华颖
课程类型：必修
学时/学分：1学期

## Undergraduate Program 4th Year
### THESIS PROJECT: SPECIALIZED STUDY ON DIGITAL CONSTRUCTION TECHNOLOGIES · TONG Ziyu, ZHONG Huaying
Type: Obligatory
Study Period and Credits: 1 term

课题内容
校园数字化搭建
设计内容
南京大学拟在校园内选择某处闲置的场地或室外台阶搭建一个具有使用功能的构筑物，如休息亭、车棚、雨棚等，要求使用数字化设计和建造手段，在满足使用需求的同时，丰富校园景观环境。设定构筑物的使用功能，并顺应环境的要求。同时，构筑物被分为A、B两部分，应用不同的设计参数、算法和构筑方式，两部分延伸相交为一个整体。最后利用数控加工手段，完成1∶1原型的建造。
 1. 基地。用地位于校园内某处坡地或台地、台阶。
 2. 功能。服务校园生活。
 3. 类型。A、B两组遮盖物具有不同形态，同时交汇融合为一个整体。
 4. 材料。满足预算和现有加工条件的限制。
 5. 建造。满足自行搭建的要求。

### Subject Content
Digital Design and Fabrication on Campus
### Design Content
Nanjing University is planning to build a structure with usage functions (such as a pavilion, shelter for bicycles and canopy) at an idle place or outdoor stairs on the campus using digital design and construction means to decorate the campus landscape. The function of the structure should be defined and meet the environmental requirements. Meanwhile, the structure should be divided into Part A and Part B, with both parts employing different design parameters, algorithms and construction methods and intersecting with each other to form a whole. Finally, a complete 1∶1 prototype will be constructed with CNC processing technology.
1. Land: A hillside field, terrace or steps.
2. Function: To support campus life.
3. Type: Part A and Part B employ different forms but intersect with each other to form a whole.
4. Materials: Meeting the budget requirements and existing processing conditions.
5. Construction: Meeting the requirements of self-construction.

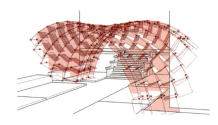

本科四年级
## 毕业设计
周凌　胡友培
课程类型：必修
学时／学分：1学期

Undergraduate Program 4th Year
## THESIS PROJECT · ZHOU Ling, HU Youpei
Type: Required Course
Study Period and Credits: 1 term

**课题内容**
　　安徽祁门洪家大屋及周边地块保护更新规划设计
**教学目标**
　　中国传统村镇正在消失，其速度与中国城市化速度成正比，传统农耕文化、传统手工艺、传统价值观处于消失弱化的边缘。本课题着重研究中国传统村镇保护更新的议题。学生通过调研和规划设计，了解传统村镇与传统建筑文化，学习规划知识，训练建筑设计技巧。
**研究主题**
　　1. 村落自然与人文环境；
　　2. 民居类型；
　　3. 保护规划；
　　4. 改造策划；
　　5. 技术与建造。
**设计内容**
　　以洪家大屋及周边地块保护更新规划设计为例，通过调研测绘访问的方式，对传统村镇进行基础研究，在研究基础上，提出整体保护和更新的概念规划。
　　通过实地调研，掌握古建筑知识与制图技巧；通过案例研究学习方法，通过讨论梳理思路，最终掌握规划与设计中分析图、平立剖图、模型等表达，完成规划与建筑设计深度的图纸与研究报告。图纸表现方式和比例自定。

**Subject Content**
Planning and Design for the Protection and Reconstruction of the Hongjia Great House and Surrounding Plots in Qimen County, Anhui Province
**Training Objective**
In China, traditional villages and small towns are disappearing proportionally with China's urbanization. At the same time, the traditional farming culture, handicraft and values are also going to disappear or fade. This topic focuses on the protection and reconstruction of Chinese traditional villages and small towns, and the students will come to understand the traditional villages and small towns and their architectural culture, learn planning knowledge and practice architectural design skills through research and planning and design.
**Research Subject**
Context and culture, typology of traditional buildings, master planning for preservation, program and functions, techniques and construction
**Design Content**
Take the "Planning and Design for the Protection and Reconstruction of the Hongjia Great House and Surrounding Plots" as an example, conduct a fundamental study on traditional villages and towns through research, mapping and interviews, and then propose a conceptual plan for overall protection and reconstruction.
Students should grasp ancient architectural knowledge and mapping skills through field research, master the expression skills in planning and design including analysis charts, floor plans, elevation drawings, sectional drawings and models through case study and discussion, and complete the drawing and study report for planning and architectural design. The expression mode and scale of the drawing can be self-defined.

研究生一年级
## 建筑设计研究（一）基本设计
傅筱
课程类型：必修
学时／学分：40学时／2学分

Graduate Program 1st Year
## DESIGN STUDIO 1: BASIC DESIGN · FU Xiao
Type: Required Course
Study Period and Credits: 40 hours/2 credits

**课题内容**
　　南大附属九班幼儿园设计
**教学目标**
　　课程从"空间"、"场所"与"建造"等建筑的基本问题出发，通过幼儿园设计，着重训练学生对建筑与基地、空间与行为等关系的认知，从而加深对建筑设计过程与设计方法的基本认识。
**研究主题**
　　建筑形态与周边环境、空间构成与行为模式。
**设计内容**
　　在老城区设计1个9班幼儿园。

**Subject Content**
Kindergarten Design
**Course Objective**
The course starts from some basic items, such as space, site and construction, and focuses on enhancing students' awareness of relations between building and consturction sites, spaces to behaviors via kindergarten design, in order to help students get a basic understanding on building design process and methods.
**Research Subject**
Structure of building form to surrounding environment, spaces and behavior model.
**Design Content**
Design a 9 classes kindergarten in the old city area in Erdos.

研究生一年级
## 建筑设计研究（二）基本设计
张雷
课程类型：必修
学时 / 学分：40 学时 / 2 学分

Graduate Program 1st Year
## DESIGN STUDIO 2: BASIC DESIGN · ZHANG Lei
Type: Required Course
Study Period and Credits: 40 hours/2 credits

**课题内容**
南京老城南传统院落改建研究
**教学目标**
课程从"环境"、"空间"、"场所"与"建造"等基本的建筑问题出发，通过南京老城南城市肌理和建筑类型的分析，以及功能置换后的使用空间重新划分研究，从建筑与基地、空间与活动、材料与实施等关系入手，强化设计问题的分析，强调准确的专业性表达。通过基本设计训练，达到对建筑设计过程与方法的基本认识与理解。
**研究主题**
建筑类型、空间再划分、建筑更新、建造逻辑。
**设计内容**
对老城南升州路北侧评事街至大板巷段、评事街两侧进行调研，结合该区域改造规划，每组选择一个区域，每人选择一个院落，通过功能置换和整修改造，使其满足新的使用要求。

**Subject Content**
Renovation of the North Side of Shenzhou Road and Both Sides of Pingshi Land
**Training Objective**
Based on the basic architectural problems such as environment, space and place, and construction, the course asks students to begin with analyzing the relationship of buildings and the site, space and behavior, materials and implementation, to understand the old city fabric and later displaced function, and combine the professional expression with the analysis of the architectural problems, so as to comprehend the basic architectural design process and design method.
**Research Subject**
Architectural types, space redefine, architectural renovation, tectonic logic.
**Design Content**
Based on the investigation of the site, each group chooses one court or region to make it satisfy the new requirement through the functional replacement and spatial reognization.

---

研究生一年级
## 建筑设计研究（三）概念设计
冯路
课程类型：必修
学时 / 学分：40 学时 / 2 学分

Graduate Program 1st Year
## DESIGN STUDIO 3: CONCEPTUAL DESIGN · FENG Lu
Type: Required Course
Study Period and Credits: 40 hours/2 credits

**课题内容**
双重空间
**教学目标**
城市更新是当代中国诸多城市面临的重要问题。在过去的大拆大建模式变得不可持续时，需要探讨新的城市空间生成方式。这种新的方式不是简单的建筑形式更新，而是包含社会文化经济和政治意义的空间生产。通过本设计让学生加深理解中国当代城市的特有状况，并探索从中获得新能量的方式。
**研究主题**
新旧建筑混合使用的可能性。
**设计内容**
基地在已经基本消失的老上海张家花园，现在茂民北路以东，吴江路以南。基地现为上海中心城区较为典型的石库门建筑区，居住与小商业及零散加工业混杂共处。张家花园是老上海公共租界内第一个城市公共空间，内容丰富。设计目标是改造该区域，把消失的张家花园以某种新的方式重新投射到现实空间里。设计意图改善居住与城市更新，与此同时，试图探索不同于常规功能混合模式的空间可能性。

**Subject Content**
Double Space
**Training Objective**
Urban renewal is an important problem that many cities in China are facing. When the original mode of demolition and reconstruction became unsustainable, we need to develop a new urban space generation mode. This mode not only involves simple architectural form renewal, but also contains space generation in social, cultural, economic and political senses. Through this design task, the students will further understand the unique status of contemporary Chinese cities, and explore a method of obtaining new energy from this.
**Study Topics**
Possibility of mixed use of new and old buildings.
**Design Content**
The base is the almost disappeared Zhang's Garden of the old Shanghai, located east of North Maomin Road and South of Wujiang Road. Currently, here is the typical Shikumen Building Area in Shanghai's downtown, which mixes the functions of residence, small business and scattered processing. Zhang's Garden is the first urban public space in old Shanghai's international settlement, with abundant content. Our design objective is to reform this area so as to display the disappeared Zhang's Garden in the real world with a new method. The design intent is exploring the space possibilities different from the common function mixing mode while housing condition improvement and urban renewal.

研究生一年级
# 建筑设计研究（四）概念设计
鲁安东
课程类型：必修
学时 / 学分：40 学时 / 2 学分

Graduate Program 1st Year
## DESIGN STUDIO 4: CONCEPTUAL DESIGN · LU Andong
Type: Required Course
Study Period and Credits: 40 hours/2 credits

**课题内容**
水乡聚落设计研究·呼吸作用
**背景**
　　1. 位于长三角城市群内部缝隙地带的村庄，正在重新形成与城市的关系，因而面对"扩张"与"收缩"的双重不确定性。
　　2. 村庄的自然生长日益受到城市开发模式的干预，使得村庄原有的自维持和自调节能力衰竭。
　　3. 复杂的环境问题：极端气候、水灾、土地资源短缺、环境污染。
　　4. 村落格局的变化：从滨水型向基础设施型。
**研究主题**
　　1. 空间——呼吸作用；
　　2. 社会——再兴；
　　3. 自然——抵抗力；
　　4. 建造——再利用。
**设计内容**
金坛市儒林镇汤墅村及下辖诸自然村。
　　1. 基于"非正式"的村庄规划：利用基础设施支持呼吸作用，重新定位村庄与水的关系（水利与水害）；
　　2. 可持续的聚落设计：宅基地和自留地的调整、新农房设计、村庄加密；
　　3. 可持续的建筑设计：农房与公建——本地小机械的施工条件、可回收材料、被动式技术（厚墙）、绿色砌块和透水砖、生产性建筑（立体农业）。

**Subject Content**
Study of Design of the Waterfront Settlement · Respiration
**Background**
1. The relationship between the villages located at the gaps among the Yangtze River Delta city group and the cities is reforming; therefore, the villages are facing the uncertainty of "expansion" or "shrink".
2. The villages' growth increasingly intervened by the urban development modes, which weakens the villages' self-maintain and self-adjust capabilities.
3. Complicated environment problem: extreme weather, floods, land resource shortages and environmental pollution.
4. Change of the village pattern: from the waterfront type to infrastructure type.
**Research Subject**
1. Space-Respiration;　　2.Society-Revitalization;
3. Nature-Resilience;　　4.Construction-Reuse.
**Design Content**
Tangshu Village and other villages under the control of Rulin Town, Jintan City.
1. Based on "informal" village planning: support respiration by the infrastructure; redefine the relationship between the village and water (the advantages and disadvantages of water);
2. Sustainable settlement design: adjustment of the house sites and private plots, new rural housing design and village concentration;
3. Sustainable architectural design: rural housing and public buildings—local small machinery's construction conditions, recyclable materials, passive technology (thick walls), green blocks and permeable brick and productive buildings (stereo agriculture).

---

研究生一年级
# 建筑设计研究（五）建构设计
傅筱
课程类型：必修
学时 / 学分：40 学时 / 2 学分

Graduate Program 1st Year
## DESIGN STUDIO 5: CONSTRUCTIONAL DESIGN · FU Xiao
Type: Required Course
Study Period and Credits: 40 hours/2 credits

**课题内容**
"基本设计"的深化与发展
**教学目的**
　　1. 训练学生对设计概念与构造设计关联性的认知：
　　（1）处理好节点的基本工程技术问题。
　　A. 对自然力的抵抗与利用：保温、防水、遮阳……
　　B. 构造与施工：复杂问题简单化、建造方便性、误差问题……
　　（2）根据设计概念研究建造材料的选用和节点设计，在满足基本工程技术的前提下，重点研究超越基本工程技术问题的构造设计表达。
　　2. 训练学生对一个"完整空间形态"建造的认知：
所谓完整空间形态是指包括外墙（从屋顶到基础）、设备、装修所构成的空间形态，通过完整空间形态设计，让学生建立构造设计的整体意识。
**设计内容**
以基本计案例为基础，选取1—2个主要设计概念进行深化设计，要求达到节点大样表达深度。

**Subject Content**
Deepin development of the Projects of Basic Design
**Training Objective**
1. Make students understanding the relationship between design concept and detail of construction.
(1) Solute the basic technical problems of detail.
a. Resistance and application of natural elements: heat preservation, water resistance, sun shading, etc.
b. Construction and building: simplification of complicated problem, conveniences of construction, deviation, etc.
(2) Study the selection of materials and design of details according to the design concepts, especially the presentation of details beyond the basic technical problem.
2. Make students understanding the construction of an integrated space.
The integrated space includes the envelope (from roof to foundation), facilities and decoration. Students will get the integrated awareness of detail design via the design practice of integrated space.
**Design Content**
Deepeningly design one or two concepts from the projects of Basic Design to the detail level.

研究生一年级
建筑设计研究（六）建构设计
郭屹民
课程类型：必修
学时 / 学分：40 学时 / 2 学分

Graduate Program 1st Year
**DESIGN STUDIO 6:
CONSTRUCTIONAL DESIGN**
GUO Yimin
Type: Required Course
Study Period and Credits: 40 hours/2 credits

课题内容
南京大学校园临时展廊
教学目标
1. 掌握结构设计基础知识，并会进行结构分析和结构设计。
2. 了解结构的材料与建造，并会通过材料和建造进行建筑构造设计。
3. 了解结构设计与功能、场地的关系，并会进行与功能相关的建筑设计。
设计内容
为了在南大校区向兄弟院系展示南大建筑的学科特色和教学成绩，拟在南大校区北苑范围内建设一处临时展廊，供南大建筑城规学院建筑系本科毕业设计和答辩所用。展廊为临时建筑，要求满足搭建时间1个月的强度要求。并同时考虑材料、组织、建造、拆除及其对于场地的影响。通过抽象结构力学→建筑物质形式的生成方法，将结构与空间创造一体化。

**Subject Content**
The Temporary Exhibition Gallery in Nanjing University
**Training Objective**
1. To grasp the basic knowledge of structural design, and to be able to conduct structural analysis and design.
2. To understand the materials and construction of the structure, and to be able to conduct building structure design with the materials and through construction.
3. To understand the relationships between structural design and the functions as well as sites, and to be able to conduct function-related architectural design.
**Design Content**
For the purpose of demonstrating the subject characteristics and training achievements of the School of Architecture and Urban Planning, Nanjing University, we are planning to build up a temporary exhibition gallery in Beiyuan area of Nanjing University for the seniors of the Department of Architecture to demonstrate their thesis projects and conduct thesis defense. The exhibition gallery is a temporary building, which is required to build up within one month and upon consideration of the materials, organization, construction, remove and its effect on the site. Structural design and space creation should be integrated through the generation method that from the abstract structural mechanics to building material form.

研究生一年级
建筑设计研究（七）城市设计
丁沃沃
课程类型：必修
学时 / 学分：40 学时 / 2 学分

Graduate Program 1st Year
**DESIGN STUDIO 7: URBAN DESIGN** · DING Wowo
Type: Required Course
Study Period and Credits: 40 hours/2 credits

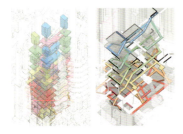

课题内容
城市空间设计
教学目标
城市化进程改变国土的自然地貌和人的生存空间。事实上，先行完成城市化的西方发达国家尤其是欧洲早已意识到平面扩张城市的恶果，转而在空间上寻求突破，由此引发了丰富多样的城市设计方法的探索。集约化城市的要义之一是空间的集约，然而高密度空间引发的问题往往制约了集约化城市的实现。本课题拟通过城市空间设计实验了解城市物质空间的本质和效能，初步掌握城市空间分配和城市建筑之间的关系。此外，通过城市空间设计练习进一步深化空间设计的技能和方法。
研究主题
以"层"与"界"作为操作元素，研究多维城市设计的方法，探索"多层化"的城市空间与城市地块指标体系之间的关系。
设计内容
1. 以南京新街口中心地块作为设计实验的场所，通过实地调研、案例研习、设计分析和试验，探讨高效城市空间的建构方法和内涵。
2. 观察与发现城市物质空间内的运作现象，基于建筑学图示方法，从空间意向出发建构城市物质空间的层级与界面，由"空间意向""扫描"出城市空间的"层"与"界"。

**Subject Content**
From Volume to Space
**Traning Objective**
Chinese traditional villages and towns are fading away in direct proportion to Chinese urbanization speed, and the traditional farming culture, traditional handicraft and traditional values are on the verge of disappearing and weakening. The course focuses on studying the topic about Chinese traditional village and town protection and updating. Taking Hong Family's big house in Qimen, Anhui and surrounding land protection and updating plan for example, it makes basic research on traditional villages and towns and puts forward the concept planning of overall protection and updating on the research basis. Students understand the traditional villages and towns and traditional building culture through survey and planning design, learn planning knowledge and practice building design skills.
**Research Subject**
Study the method of multi-dimensional urban design with "layer" and "interface" as operation elements and explore the relationship between "multi-layer" urban space and urban plot indicator system.
**Design Content**
1. Explore efficient urban space construction methods and connotation through field survey, case study, design analysis and experiment with Nanjing Xin Jiekou central plot as the design experiment place.
2. Observe and discover the operation phenomenon in the urban material space, start from the space intention and build thelayer and interface of urban material space on the basis of architectural graphic method, and "scan" the "layer" and "interface" of the urban space from "space intention".

研究生一年级
# 建筑设计研究（二）城市设计
鲁安东
课程类型：必修
学时 / 学分：40 学时 / 2 学分

Graduate Program 1st Year
**DESIGN STUDIO2: URBAN DESIGN** · LU Ando
Type: Required Course
Study Period and Credits: 40 hours/2 credits

**课题内容**
　　分割线——城墙下的城
**教学目标**
　　城墙是南京城市空间的边界条件，在形态和生态上导致了多样的微观环境。本课程将系统地研究南京城墙地带及其对空间的影响，并选择特征地块进行城市设计。
**设计内容**
　　本课程将介绍城市研究的工作方法，帮助设计师来理解、分析和设计城市。本课程将包括三个部分：
　　1. "城市图绘"：我们将通过对城墙地带的详细观察、分析和图示来理解这一边缘地带的限制和机会。研究的重心在于空间的历时性与层积效应。
　　2. "生态分析"：我们将使用分析技术来更好地理解城市地块的生态环境，探讨城市发展的多种资源，寻找不同的发展模式，在此基础上建立一个城市"工具盒"。
　　3. "城市设计"：城市设计的目的重新定义边界条件，在城墙地带整体系统中建构新的微观形态和生态。

**Subject Content**
Demarcation Line: City beyond Walls
**Training Objectives**
The city wall is the boundary condition of Nanjing's urban space, which causes diverse micro-environments in the morphological and ecological terms. This course will conduct a systematic study on Nanjing's city wall area and its effect on the space, and select a featured plot for urban design.
**Design Content**
This course will introduce the methods of urban study to help the architects understand, analyze and design the city. The course consists of the follow three parts:
1. Urban Mapping: We will come to understand this border area's constraints and opportunities through detailed observation, analysis and illustration on it. The study will focus on the diachronism and laminated effect of the space.
2. Ecological Analysis: We will further understand the urban plot's ecological environment with the analysis technology, discuss the various resources for urban development, seek for different development modes, and set up a "tool box" for the city as a result.
3. Urban Design: We aim at redefining the boundary condition and establishing new morphology and ecology in the overall system of the city wall area.

建筑理论课程
ARCHITECTURAL THEORY COURSES

本科二年级
建筑导论・赵辰等
课程类型：必修
学时/学分：36学时/2学分
Undergraduate Program 2nd Year
**INTRODUCTION TO ARCHITECTURE** • ZHAO Chen, etc.
Type: Required Course
Study Period and Credits:32 hours / 2 credits

课程内容
1. 建筑学的基本定义
第一讲： 建筑与设计/赵辰
第二讲： 建筑与城市/丁沃沃
第三讲： 建筑与生活/张雷
2. 建筑的基本构成
（1）建筑的物质构成
第四讲： 建筑的物质环境/赵辰
第五讲： 建筑与节能技术/秦孟昊
第六讲： 建筑与生态环境/吴蔚
第七讲： 建筑与建造技术/冯金龙
（2）建筑的文化构成
第八讲： 建筑与人文、艺术、审美/赵辰
第九讲： 建筑与环境景观/华晓宁
第十讲： 城市肌体/胡友培
第十一讲： 建筑与身体经验/鲁安东
（3）建筑师职业与建筑学术
第十二讲： 建筑与表现/赵辰
第十三讲： 建筑与几何形态/周凌
第十四讲： 建筑与数字技术/吉国华
第十五讲： 城市与数字技术/童滋雨
第十六讲： 建筑师的职业技能与社会责任/傅筱

Course Content
I Preliminary of architecture
 1. Architecture and design / ZHAO Chen
 2. Architecture and urbanization / DING Wowo
 3. Architecture and life / ZHANG Lei
II Basic attribute of architecture
II-1 Physical attribute
 4. Physical environment of architecture / ZHAO Chen
 5. Architecture and energy saving / QIN Menghao
 6. Architecture and ecological environment / WU Wei
 7. Architecture and construction technology / FENG Jinlong
II-2 Cultural attribute
 8. Architecture and civilization, arts, aesthetic / ZHAO Chen
 9. Architecture and landscaping environment / HUA Xiaoning
 10. Urban tissue / HU Youpei
 11. Architecture and body / LU Andong
II-3 Architect: profession and academy
 12. Architecture and presentation / ZHAO Chen
 13. Architecture and geometrical form / ZHOU Ling
 14. Architectural and digital technology / JI Guohua
 15. Urban and digital technology / TONG Ziyu
 16. Architect's professional technique and responsibility / FU Xiao

---

本科三年级
建筑设计基础原理・周凌
课程类型：必修
学时/学分：36学时/2学分

Undergraduate Program 3rd Year
**BASIC THEORY OF ARCHITECTURAL DESIGN**
• ZHOU Ling
Type: Required Course
Study Period and Credits:36 hours / 2 credits

教学目标
　　本课程是建筑学专业本科生的专业基础理论课程。本课程的任务主要是介绍建筑设计中形式与类型的基本原理。形式原理包含历史上各个时期的设计原则，类型原理讨论不同类型建筑的设计原理。
课程内容
　　1. 形式与类型概述
　　2. 古典建筑形式语言
　　3. 现代建筑形式语言
　　4. 当代建筑形式语言
　　5. 类型设计
　　6. 材料与建造
　　7. 技术与规范
　　8. 课程总结
课程要求
　　1. 讲授大纲的重点内容；
　　2. 通过分析实例启迪学生的思维，加深学生对有关理论及其应用、工程实例等内容的理解；
　　3. 通过对实例的讨论，引导学生运用所学的专业理论知识，分析、解决实际问题。

Training Objective
This course is a basic theory course for the undergraduate students of architecture. The main purpose of this course is to introduce the basic principles of the form and type in architectural design. Form theory contains design principles in various periods of history; type theory discusses the design principles of different types of building.
Course Content
1. Overview of forms and types
2. Classical architecture form language
3. Modern architecture form language
4. Contemporary architecture form language
5. Type design
6. Materials and construction
7. Technology and specification
8. Course summary
Course Requirement
1. Teach the key elements of the outline;
2. Enlighten students' thinking and enhance students' understanding of the theories, its applications and project examples through analyzing examples;
3. Guide students using the professional knowledge to analysis and solve practical problems through the discussion of examples.

---

本科三年级
居住建筑设计与居住区规划原理・冷天 刘铨
课程类型：必修
学时/学分：36学时/2学分

Undergraduate Program 3rd Year
**THEORY OF HOUSING DESIGN AND RESIDENYTIAL PLANNING** • LENG Tian, LIU Quan
Type: Required Course
Study Period and Credits:36 hours / 2 credits

课程内容
第一讲： 课程概述
第二讲： 居住建筑的演变
第三讲： 套型空间的设计
第四讲： 套型空间的组合与单体设计（一）
第五讲： 套型空间的组合与单体设计（二）
第六讲： 居住建筑的结构、设备与施工
第七讲： 专题讲座：住宅的适应性，支撑体住宅
第八讲： 城市规划理论概述
第九讲： 现代居住区规划的发展历程
第十讲： 居住区的空间组织
第十一讲： 居住区的道路交通系统规划与设计
第十二讲： 居住区的绿地景观系统规划与设计
第十三讲： 居住区公共设施规划、竖向设计与管线综合
第十四讲： 专题讲座：住宅产品开发
第十五讲： 专题讲座：住宅产品设计实践
第十六讲： 课程总结，考试答疑

Course Content
Lect. 1: Introduction of the course
Lect. 2: Development of residential building
Lect. 3: Design of dwelling space
Lect. 4: Dwelling space arrangement and residential building design (1)
Lect. 5: Dwelling space arrangement and residential building design (2)
Lect. 6: Structure, detail, facility and construction of residential buildings
Lect. 7: Adapt ability of residential building, supporting house
Lect. 8: Introduction of the theories of urban planning
Lect. 9: History of modern residential planning
Lect. 10: Organization of residential space
Lect. 11: Traffic system planning and design of residential area
Lect. 12: Landscape planning and design of residential area
Lect. 13: Public facilities and infrastructure system
Lect. 14: Real estate development
Lect. 15: The practice of residential planning and housing design
Lect. 16: Summary, question of the test

**Graduate Program 1st Year**
**PRELIMINARIES IN MODERN ARCHITECTURAL DESIGN** · ZHANG Lei
Type: Required Course
Study Period and Credits: 18 hours/1 credit

**Course Content**
1. Transition of the modern thoughts of design
2. Arrangement of basic space
3. Abstraction and reversion of architectural types
4. Material application and constructional issues
5. Formation and significance of sites
6. Nowaday working principles and strategies

Architecture can be abstracted to the most fundamental state of space enclosure, so as to confront all the basic applicable problems which must be resolved. The most reasonable and direct mode of space arrangement and construction shall be applied; ordinary materials and universal methods shall be used as the countermeasures to the complicated application requirement. These are the basic principles on which an architecture design institution shall focus.

**Graduate Program 1st Year**
**METHODOLOGY OF MODERN ARCHITECTURAL DESIGN** · DING Wowo
Type: Required Course
Study Period and Credits: 18 hours/1 credit

**Course Content**
Along the main line of architectural history, this course discussed the evolution of architectural design motivation ideas and methodology. Due to different concepts between the Chinese architecture and Western architecture Matters. The way for analyzing and exploring has to be studied. By analyzing the logic relationship of architectural form language, the geometrical significance of architectural form language is explored. Finally, within the context of urban form and space, the significance of architectural autonomy has been discussed.

1. Introdution
2. Tradition of western architecture
3. Meaning of architecture in China
4. History and modernity
5. Modern architectural ideology and its dilemma
6. Exploration for architectural form language
7. Re-thinking and return to reason
8. Conclusion

**Graduate Program 1st Year**
**CINEMATIC ARCHITECTURE** · LU Andong
Type: Elective Course
Study Period and Credits: 36 hours/2 credits

**Course Content**
Film, in this course, is seen as a distinctive medium for the perception of space and the communication of thoughts. We shall learn how to use the unique narrative medium of film to conduct a microscopic study on architecture and urbanism. This course will teach the students of a new way of seeing and knowing architecture. Its purpose is not only to teach theories of urbanism and techniques of filmmaking, but also to teach the students, through a complete case study, of a cinematic (visceral and non-abstract) way of thinking, analyzing and presenting ideas. This course is composed of a series of 4-hour sessions, which gradually lead the students to undertake their own research project and to produce a cinematic essay on a case study of their own choice. The teaching of this course will be conducted in three forms: the lectures will introduce the students to cinematic ways of seeing and understanding architecture; the tutorials will introduce the students to some basic cinematic techniques, including continuity editing, cinematography, storyboard, shooting script, and post-production; the seminars will review and discuss the students' works in several stages.

城市理论课程
URBAN THEORY COURSES

本科四年级
城市设计及其理论・丁沃沃 胡友培
课程类型：必修
学时/学分：36学时/2学分

Undergraduate Program 4th Year
**THEORY OF URBAN DESIGN** • DING Wowo, HU Youpei
Type: Required Course
Study Period and Credits: 36 hours / 2 credits

课程内容
  第一讲 课程概述
  第二讲 城市设计技术术语：城市规划相关术语；城市形态相关术语；城市交通相关术语；消防相关术语
  第三讲 城市设计方法 —— 文本分析：城市设计上位规划；城市设计相关文献；文献分析方法
  第四讲 城市设计方法 —— 数据分析：人口数据分析与配置；交通流量数据分析；功能分配数据分析；视线与高度数据分析；城市空间数据模型的建构
  第五讲 城市设计方法 —— 城市肌理分类：城市肌理分类概述；肌理形态与建筑容量；肌理形态与开放空间；肌理形态与交通流量；城市绿地指标体系
  第六讲 城市设计方法 —— 城市路网组织：城市道路结构与交通结构概述；城市路网与城市功能；城市路网与城市空间；城市路网与市政设施；城市道路断面设计
  第七讲 城市设计方法 —— 城市设计表现：城市设计分析图；城市设计概念表达；城市设计成果解析图；城市设计地块深化设计表达；城市设计空间表达
  第八讲 城市设计的历史与理论：城市设计的历史意义；城市设计理论的内涵
  第九讲 城市路网形态：路网形态的类型和结构；路网形态与肌理；路网形态的变迁
  第十讲 城市空间：城市空间的类型；城市空间结构；城市空间形态；城市空间形态的变迁
  第十一讲 城市形态学：英国学派；意大利学派；法国学派；空间句法
  第十二讲 城市形态的物理环境：城市形态与物理环境；城市形态与环境研究；城市形态与环境测研；城市形态与环境操作
  第十三讲 景观都市主义：景观都市主义的理论、操作和范例
  第十四讲 城市自组织现象及其研究：城市自组织现象的魅力与问题；城市自组织系统研究方法；典型自组织现象案例研究
  第十五讲 建筑学图式理论与方法：图式理论的研究；建筑学图式的概念；图式理论的应用；作为设计工具的图式；当代城市语境中的建筑学图式理论探索
  第十六讲 课程总结

Course Content
Lect. 1. Introduction
Lect. 2. Technical terms: terms of urban planning, urban morphology, urban traffic and fire protection
Lect. 3. Urban design methods — documents analysis: urban planning and policies; relative documents; document analysis techniques and skills
Lect. 4. Urban design methods — data analysis: data analysis of demography, traffic flow, public facilities distribution, visual and building height; modelling urban spatial data
Lect. 5. Urban design methods — classification of urban fabrics: introduction of urban fabrics; urban fabrics and floor area ratio; urban fabrics and open space; urban fabrics and traffic flow; criteria system of urban green space
Lect. 6. Urban design methods — organization of urban road network: introduction; urban road network and urban function; urban road network and urban space; urban road network and civic facilities; design of urban road section
Lect. 7. Urban design methods — representation skills of urban Design: mapping and analysis; conceptual diagram; analytical representation of urban design; representation of detail design; spatial representation of urban design
Lect. 8. Brief history and theories of urban design: historical meaning of urban design; connotation of urban design theories
Lect. 9. Form of urban road network: typology, structure and evolution of road network; road network and urban fabrics
Lect. 10. Urban space: typology, structure, morphology and evolution of urban space
Lect. 11. Urban morphology: Cozen School; Italian School; French School; Space Syntax Theory
Lect. 12. Physical environment of urban forms: urban forms and physical environment; environmental study; environmental evaluation and environmental operations
Lect. 13. Landscape urbanism: ideas, theories, operations and examples of landscape urbanism
Lect. 14. Researches on the phenomena of the urban self-organization: charms and problems of urban self-organization phenomena; research methodology on urban self-organization phenomena; case studies of urban self-organization phenomena
Lect. 15. Theory and method of architectural diagram: theoretical study on diagrams; concepts of architectural diagrams; application of diagram theory; diagrams as design tools; theoretical research of architectural diagrams in contemporary urban context
Lect. 16. Summary

研究生一年级
城市形态研究・丁沃沃 赵辰 萧红颜
课程类型：必修
学时/学分：36学时/2学分

Graduate Program 1st Year
**URBAN MORPHOLOGY** • DING Wowo, ZHAO Chen, XIAO Hongyan
Type: Required Course
Study Period and Credits: 36 hours / 2 credits

课程要求
  1．要求学生基于对历史性城市形态的认知分析，加深对中西方城市理论与历史的理解。
  2．要求学生基于历史性城市地段的形态分析，提高对中西方城市空间特质及相关理论的认知能力。
课程内容
  第一周：序言 概念、方法及成果
  第二周：讲座1 城市形态认知的历史基础 —— 营造观念与技术传承
  第三周：讲座2 城市形态认知的历史基础 —— 图文并置与意象构建
  第四周：讲座3 城市形态认知的理论基础 —— 价值判断与空间生产
  第五周：讲座4 城市形态认知的理论基础 —— 勾沉呈现与特征形塑
  第六周：讲座5 历史城市的肌理研究
  第七周：讲座6 整体与局部 —— 建筑与城市
  第八周：讨论
  第九周：讲座7 城市化与城市形态
  第十周：讲座8 城市乌托邦
  第十一周：讲座9 走出乌托邦
  第十二周：讲座10 重新认识城市
  第十三周：讲座11 城市设计背景
  第十四周：讲座12 城市设计实践
  第十五周：讲座13 城市设计理论
  第十六周：讲评

Course Requirement
1. Deepen the understanding of Chinese and Western urban theories and histories based on the cognition and analysis of historical urban form.
2. Improve the cognitive abilities of the characteristics and theories of Chinese and Western urban space based on the morphological analysis of the historical urban sites.
Course Content
Week 1. Preface — concepts, methods and results
Week 2. Lect. 1 Historical basis of urban form cognition — Developing concepts and passing of technologies
Week 3. Lect. 2 Historical basis of urban form cognition — Apposition of pictures and text and imago construction
Week 4. Lect. 3 Theoretical basis of urban form cognition — Value judgement and space production
Week 5. Lect. 4 Theoretical basis of urban form cognition — History representation and feature shaping
Week 6. Lect. 5 Study on the grain of historical cities
Week 7. Lect. 6 Whole and part: Architecture and urban
Week 8. Discussion
Week 9. Lect. 7 Urbanization and urban form
Week 10. Lect. 8 Urban Utopia
Week 11. Lect. 9 Walk out of Utopia
Week 12. Lect. 10 Have a new look of the city
Week 13. Lect. 11 Background of urban design
Week 14. Lect. 12 Practice of urban design
Week 15. Lect. 13 Theory of urban design
Week 16. Discussions

本科四年级
景观规划设计及其理论·尹航
课程类型：选修
学时/学分：36学时/2学分

Undergraduate Program 4th Year
**LANDSCAPE PALNNING DESIGN AND THEORY**
• YIN Hang
Type: Elective Course
Study Period and Credits: 36 hours / 2 credits

课程介绍
　　景观规划设计的对象包括所有的室外环境，景观与建筑的关系往往是紧密而互相影响的，这种关系在城市中尤为明显。景观规划设计及理论课程希望从景观设计理念、场地设计技术和建筑周边环境塑造等方面开展课程的教学，为建筑学本科生建立更加全面的景观知识体系，并且完善建筑学本科生在建筑场地设计、总平面规划与城市设计等方面的设计能力。
　　本课程主要从三个方面展开：一是理念与历史：以历史的视角介绍景观学科的发展过程，让学生对景观学科有一个宏观的了解，初步理解景观设计理念的发展；二是场地与文脉：通过阐述景观规划设计与周边自然环境、地理位置、历史文脉和方案可持续性的关系，建立场地与文脉的设计思维；三是景观与建筑：通过设计方法授课、先例分析作业等方式让学生增强建筑的环境意识，了解建筑的场地设计的影响因素、一般步骤与设计方法，并通过与"建筑设计6"和"建筑设计7"的设计任务书相配合的同步课程设计训练来加强学生景观规划设计的能力。

**Course Description**
The object of landscape planning design includes all outdoor environments; the relationship between landscape and building is often close and interactive, which is especially obvious in a city. This course expects to carry out teaching from perspective of landscape design concept, site design technology, building's peripheral environment creation, etc. to establish a more comprehensive landscape knowledge system for the undergraduate students of architecture, and perfect their design ability in building site design, master plane planning and urban design and so on.
This course includes three aspects:
1. Concept and history;
2. Site and context;
3. Landscape and building.

---

本科四年级
东西方园林·许浩
课程类型：选修
学时/学分：36学时/2学分

Undergraduate Program 4th Year
**GARDEN OF EAST AND WEST** • XU Hao
Type: Elective Course
Study Period and Credits: 36 hours / 2 credits

课程介绍
　　帮助学生系统掌握园林、绿地的基本概念、理论和研究方法，尤其了解园林艺术的发展脉络，侧重各个流派如日式园林、江南私家园林、皇家园林、规则式园林、自由式园林、伊斯兰园林的不同特征和关系，使得学生能够从社会背景、环境等方面解读园林的发展特征，并能够开展一定的评价。

**Course Description**
Help students systematically master the basic concepts, theories and research methods of gardens and greenbelts, especially understand the evolution of gardening, emphasizing the different features and relationships of various genres, such as Japanese gardens, private gardens by the south of Yangtze River, royal gardens, rule-style gardens, free style gardens and Islamic gardens; enable students to interpret the characteristics of garden development from the point of view of social backgrounds, environment, etc. Furthermore to do the evaluation in depth.

---

研究生一年级
景观规划进展·许浩
课程类型：选修
学时/学分：18学时/1学分

Graduate Program 1st Year
**LANDSCAPE PLANNING PROGRESS** • XU Hao
Type: Elective Course
Study Period and Credits: 18 hours / 1 credit

课程介绍
　　生态规划是景观规划的核心内容之一。本课程总结了生态系统、生态保护和生态修复的基本概念。大规模生态保护的基本途径是国家公园体系，而生态修复则是通过人为干涉对破损环境的恢复。本课程介绍了国家公园的价值、分类与成就，并通过具体案例论述了欧洲、澳洲景观设计过程中生态修复的做法。

**Course Description**
Ecological planning is one of the core contents of landscape planning. This course summarizes the basic concepts of ecological systems, ecological protection and ecological restoration. The basic channel of large-scale ecological protection is the national park system, while ecological restoration is to restore the damaged environment by means of human intervention. This course introduces the values, classification and achievements of national parks, and discusses the practices of ecological restoration in the process of landscape design in Europe and Australia.

---

研究生一年级
景观都市主义理论与方法·华晓宁
课程类型：选修
学时/学分：18学时/1学分

Graduate Program 1st Year
**THEORY AND METHOD OF LANDSCAPE URBANISM**
• HUA Xiaoning
Type: Elective Course
Study Period and Credits: 18 hours / 1 credit

课程介绍
　　本课程介绍了景观都市主义思想产生的背景、缘起及其主要理论观点，并结合实例，重点分析了其在不同的场址和任务导向下发展起来的多样化的实践策略和操作性工具。通过这些内容的讲授，本课程的最终目的是拓宽学生的视野，引导学生改变既往的思维定式，以新的学科交叉整合的思路，分析和解决当代城市问题。

课程内容
　　第一讲　导论——当代城市与景观媒介
　　第二讲　生态过程与景观修复
　　第三讲　基础设施与景观嫁接
　　第四讲　嵌入与缝合
　　第五讲　水平性与都市表面
　　第六讲　城市图绘与图解
　　第七讲　AA景观都市主义——原型方法
　　第八讲　总结与作业

**Course Description**
The course introduces the backgrounds, the generation and the main theoretical opinions of landscape urbanism. With a series of instances, it particularly analyses the various practical strategies and operational techniques guided by various sites and projects. With all these contents, the aim of the course is to widen the students' field of vision, change their habitual thinking and suggest them to analyze and solve contemporary urban problems using the new ideas of the intersection and integration of different disciplines.

**Course Content**
Lect. 1: Introduction — contemporary cities and landscape medium
Lect. 2: Ecological process and landscape recovering
Lect. 3: Infrastructure and landscape engrafting
Lect. 4: Embedment and oversewing
Lect. 5: Horizontality and urban surface
Lect. 6: Urban mapping and diagram
Lect. 7: AA Landscape Urbanism — archetypical method
Lect. 8: Conclusion and assignment

历史理论课程
HISTORY THEORY COURSES

Undergraduate Program 2nd Year
**HISTORY OF CHINESE ARCHITECTURE (ANCIENT)**
• XIAO Hongyan
Type: Required Course
Study Period and Credits: 36 hours / 2 credits

**Training Objective**
Recognize the thinking characteristics and technology selection of China's traditional construction; develop students' awareness of understanding history and analyzing problems.
**Course Content**
Use a teaching method different from traditional teaching, concurrently emphasize special lecturing and type analysis as well as the problem-based teaching under diversified visions, attempting to realize the integrated effect of cognition and learnt theory.

---

Undergraduate Program 2nd Year
**HISTORY OF WESTERN ARCHITECTURE (ANCIENT)** •
HU Heng
Type: Required Course
Study Period and Credits: 36 hours / 2 credits

**Training Objective**
This course seeks to give an overall outline of Western architectural history, so that the students may have an in-depth understanding of the structural transition (different styles of evolution) of Western architectural history in the past 2000 years. This course hopes that students can understand the close association between the development of architectural history and the development of human civilization.
**Course Content**
1. Generality  2. Greek Architectures   3. Roman Architectures
4. The Middle Ages Architectures
5. The Middle Ages Architectures in Italy  6. Renaissance
7. Baroque   8. American Cities   9. Nordic Romanticism
10. Catalonian Architectures   11. Avant-Garde
12. German Manufacturing Alliance and Bauhaus
13. Soviet Architecture and Cities   14. 1960's Architectures
15. 1970's Architectures   16. Answer Questions

---

Undergraduate Program 3rd Year
**HISTORY OF WESTERN ARCHITECTURE (MODERN)**
• HU Heng
Type: Required Course
Study Period and Credits: 36 hours / 2 credits

**Training Objective**
This course seeks to make a detailed explanation to the works of 7 representative architects in the Renaissance period and 5 important contemporary architects in a special way. This course will try to reorganize all works of these important architects, so that the students can fully grasp their design ideas, theoretical subject and their particular relevance with the era and significance in the architectural history.
**Course Content**
1. Brunelleschi  2. Alberti  3. Bramante
4. Michelangelo(1)  5. Michelangelo(2)
6. Romano  7. Sansovino  8. Paratio(1)  9. Paratio(2)
10. Wright  11. Mies  12. Le Corbusier(1)  13. Le Corbusier(2)
14. Hejduk  15. Kazuyo Sejima
16. Answer Questions

---

Undergraduate Program 3rd Year
**HISTORY OF CHINESE ARCHITECTURE (MODERN)**
• ZHAO Chen
Type: Required Course
Study Period and Credits: 36 hours / 2 credits

**Course Description**
As the history and theory course for undergraduate students of Architecture, this course is part of the teaching of History of Chinese Architecture. Based on the earlier studying of Chinese and Western history of ancient architecture, understand the evolution progress of Chinese society's entry into modern times and even contemporary age.
Based on the comparison of Chinese and Western building culture, establish the overall understanding of China's modern and contemporary buildings. Have further understanding of the significance of China's traditional building culture's gradual evolution into one part of today's world building culture under conflict and blending with Western building culture in modern times.

Graduate Program 1st Year
**STUDY OF ARCHITECTURAL THEORY** • WANG Junyang
Type: Required Course
Study Period and Credits:18 hours / 1 credit

**Course Description**
This course is a part of teaching Western architectural history for graduate students. It mainly deals with the representative thoughts and theories in Western architectural circles, including historicism, vanguard building, critical theory, construction culture and interpretation of contemporary cities and more. Using a lot of pictures involving extensive fields including philosophy, history, art, etc., this course attempts to show the relative independence and relevance of architectural thoughts and theories under the development background of Western culture, understand the social and cultural significance owned by architectures as human activities, and inspire students' theoretical thinking and critical spirit.
**Course Content**
Lect. 1. Introduction to architectural theories
Lect. 2. Autonomous architecture
Lect. 3. Colin Rowe : the mathematics of the ideal villa and others
Lect. 4. Adolf Loos and adornment aesthetics
Lect. 5. Koolhaas and the interpretation of con-temporary cities
Lect. 6. Conscious dilemma: the reflection of modern architecture
Lect. 7. Studies in tectonic culture of Frampton
Lect. 8. Phenomenology

Graduate Program 1st Year
**ARCHITECTURAL HISTORY RESEARCH** • HU Heng
Type: Elective Course
Study Period and Credits:36 hours / 2 credits

**Training Objective**
This course has two objectives: 1. Give the students a rough understanding of the overall status of the research approaches of the Western architectural history through an overview of them. 2. Show students the approaches, point of view, concept definition, structure layout, theme settings and so on of contemporary history study through proposing the concept of contemporary history and several case studies.
**Course Content**
1. The overview of the method of architectural history(1)
2. The overview of the method of architectural history(2)
3. The overview of the method of architectural history(3)
4. Tafuri's study method of architectural history
5. The study method of contemporary history — period
6. The study method of contemporary history — hybridization
7. The study method of contemporary history — limen
8. The study method of contemporary history — opposition

Graduate Program 1st Year
**ARCHITECTURAL HISTORY RESEARCH** • XIAO Hongyan
Type: Elective Course
Study Period and Credits:36 hours / 2 credits

**Training Objective**
This course attempts to start with two areas (concept and type) to state the mutation and continuation, association and implication of basic types of the basic concept of China's traditional construction, emphasizing that the architectural history should return to the art history framework to state relevant history evidence issues and its methods.
**Course Content**
1. Corner
2. Geomantic Omen
3. Boundary
4. The Spreading and Copying
5. The Table Land of Palace
6. Pond
7. The Temple and Mausoleum
8. The Storied Building Pavilion

建筑技术课程
ARCHITECTURAL TECHNOLOGY COURSES

本科二年级
CAAD理论与实践・童滋雨
课程类型：必修
学时/学分：36学时/2学分

Undergraduate Program 2nd Year
**THEORY AND PRACTICE OF CAAD** • TONG Ziyu
Type: Required Course
Study Period and Credits: 36 hours / 2 credits

课程介绍
　　在现阶段的CAD教学中，强调了建筑设计在建筑学教学中的主干地位，将计算机技术定位于绘图工具，本课程就是帮助学生可以尽快并且熟练地掌握如何利用计算机工具进行建筑设计的表达。课程中整合了CAD知识、建筑制图知识以及建筑表现知识，将传统CAD教学中教会学生用计算机绘图的模式向教会学生用计算机绘制有形式感的建筑图的模式转变，强调准确性和表现力作为评价CAD学习的两个最重要指标。
　　本课程的具体学习内容包括：
　　1. 初步掌握AutoCAD软件和SketchUP软件的使用，能够熟练完成二维制图和三维建模的操作；
　　2. 掌握建筑制图的相关知识，包括建筑投影的基本概念，平立剖面、轴测、透视和阴影的制图方法和技巧；
　　3. 图面效果表达的技巧，包括黑白线条图和彩色图纸的表达方法和排版方法。

**Course Description**
The core position of architectural design is emphasized in the CAD course. The computer technology is defined as drawing instrument. The course helps students learn how to make architectural presentation using computer fast and expertly. The knowledge of CAD, architectural drawing and architectural presentation are integrated into the course. The traditional mode of teaching students to draw in CAD course will be transformed into teaching students to draw architectural drawing with sense of form. The precision and expression will be emphasized as two most important factors to estimate the teaching effect of CAD course.
Contents of the course include:
1. Use AutoCAD and SketchUP to achieve the 2-D drawing and 3-D modeling expertly.
2. Learn relational knowledge of architectural drawing, including basic concepts of architectural projection, drawing methods and skills of plan, elevation, section, axonometry, perspective and shadow.
3. Skills of presentation, including the methods of expression and lay out using mono and colorful drawings

---

本科三年级
建筑技术 1——结构、构造与施工・傅筱
课程类型：必修
学时/学分：36学时/2学分

Undergraduate Program 3rd Year
**ARCHITECTURAL TECHNOLOGY 1 — STRUCTURE, CONSTRUCTION AND EXECUTION** • FU Xiao
Type: Required Course
Study Period and Credits:36 hours / 2 credits

课程介绍
　　本课程是建筑学专业本科生的专业主干课程。本课程的任务主要是以建筑师的工作性质为基础，讨论一个建筑生成过程中最基本的三大技术支撑（结构、构造、施工）的原理性知识要点，以及它们在建筑实践中的相互关系。

**Course Description**
The course is a major course for the undergraduate students of architecture. The main purpose of this course is based on the nature of the architect's work, to discuss the principle knowledge points of the basic three technical supports in the process of generating construction (structure, construction, execution), and their mutual relations in the architectural practice.

---

本科三年级
建筑技术 2——建筑物理・吴蔚
课程类型：必修
学时/学分：36学时/2学分

Undergraduate Program 3rd Year
**ARCHITECTURAL TECHNOLOGY 2 — BUILDING PHYSICS** • WU Wei
Type: Required Course
Study Period and Credits:36 hours / 2 credits

课程介绍
　　本课程是针对三年级学生所设计，课程介绍了建筑热工学、建筑光学、建筑声学中的基本概念和基本原理，使学生掌握建筑的热环境、声环境、光环境的基本评估方法，以及相关的国家标准。完成学业后在此方向上能阅读相关书籍，具备在数字技术方法等相关资料的帮助下，完成一定的建筑节能设计的能力。

**Course Description**
Designed for the Grade-3 students, this course introduces the basic concepts and basic principles in architectural thermal engineering, architectural optics and architectural acoustics, so that the students can master the basic methods for the assessment of building's thermal environment, sound environment and light environment as well as the related national standards. After graduation, the students will be able to read the related books regarding these aspects, and have the ability to complete certain building energy efficiency designs with the help of the related digital techniques and methods.

---

本科三年级
建筑技术 3——建筑设备・吴蔚
课程类型：必修
学时/学分：36学时/2学分

Undergraduate 3rd Year
**ARCHITECTURAL TECHNOLOGY 3 — BUILDING EQUIPMENT** • WU Wei
Type: Required Course
Study Period and Credits:36 hours / 2 credits

课程介绍
　　本课程是针对南京大学建筑与城市规划学院本科学生三年级所设计。课程介绍了建筑给水排水系统、采暖通风与空气调节系统、电气工程的基本理论、基本知识和基本技能，使学生能熟练地阅读给水、暖通工程图，熟悉水电及消防的设计、施工规范，了解燃气供应、安全用电及建筑防火、防雷的初步知识。

**Course Description**
This course is an undergraduate class offered in the School of Architecture and Urban Planning, Nanjing University. It introduces the basic principle of the building services systems, the technique of integration amongst the building services and the building. Throughout the course, the fundamental importance to energy, ventilation, air-conditioning and comfort in buildings are highlighted.

---

研究生一年级
传热学与计算流体力学基础・郜志
课程类型：选修
学时/学分：18学时/1学分

Graduate Program 1st Year
**FUNDAMENTALS OF HEAT TRANSFER AND COMPUTATIONAL FLUID DYNAMICS** • GAO Zhi
Type: Elective Course
Study Period and Credits: 18 hours / 1 credit

课程介绍
　　本课程的主要任务是使建筑学/建筑技术学专业的学生掌握传热学和计算流体力学的基本概念和基础知识，通过课程教学，使学生熟悉传热学中导热、对流和辐射的经典理论，了解传热学和计算流体力学的实际应用和最新研究进展，为建筑能源和环境系统的计算和模拟打下坚实的理论基础。教学中尽量简化传热学和计算流体力学经典课程中复杂公式的推导过程，而着重于如何解决建筑能源与建筑环境中涉及流体流动和传热的实际应用问题。

**Course Description**
This course introduces students majoring in building science and engineering / building technology to the fundamentals of heat transfer and computational fluid dynamics (CFD). Students will study classical theories of conduction, convection and radiation heat transfers, and learn advanced research developments of heat transfer and CFD. The complex mathematics and physics equations are not emphasized. It is desirable that for real-case scenarios students will have the ability to analyze flow and heat transfer phenomena in building energy and environment systems.

研究生一年级
建筑节能与可持续发展·秦孟昊
课程类型：选修
学时/学分：18学时/1学分

Graduate Program 1st Year
ENERGY CONSERVATION AND SUSTAINABLE ARCHITECTURE · QIN Menghao
Type: Elective Course
Study Period and Credits: 18 hours / 1 credit

课程介绍
　　随着我国建筑总量的攀升和居住舒适度的提高使建筑能耗急剧上升，建筑节能就成为影响能源安全和提高能效的重要因素之一。建筑节能的关键首先是要设计"本身节能的建筑"，建筑师必须从建筑设计的最初阶段，在建筑的形体、结构、开窗方式、外墙选材等方面融入节能设计的定量分析。而这些很难通过传统建筑设计方法达到，必须依靠建筑技术、建筑设备等多学科互动协作才能完成。这已成为世界各大建筑与城市规划学院教学的一个重点。
　　本课程将采用双语教学，主要面向建筑设计专业学生讲授建筑物理、建筑技术专业关于建筑节能方面的基本理念、设计方法和模拟软件，并指导学生将这些知识互动运用到节能建筑设计的过程中，在建筑设计专业和建筑技术专业之间建立一个互动的平台，从而达到设计"绿色建筑"的目标，并为以后开展交叉学科研究、培养复合型人才奠定基础。

Course Description
With the rising of China's total number of buildings and the need for living comfort, building energy consumption is rising sharply. Building energy efficiency has become one of the key factors influencing the energy security and energy efficiency. The first key for building energy efficiency is to design "a building that conserves energy itself" and architects must carry out planning at the very beginning of building design. However, it is difficult to satisfy them by means of traditional architectural design approaches; it must be realized by interactive collaboration of diversified subjects including construction technology, construction equipment, etc. Strengthening the interaction of architectural design specialties and construction technology specialties in designing has become a key point in this course as well as in the teaching of various large architecture and urban planning colleges around the world.

研究生一年级
建筑环境学·郜志
课程类型：选修
学时/学分：18学时/1学分

Graduate Program 1st Year
FUNDAMENTALS OF BUILT ENVIRONMENT · GAO Zhi
Type: Elective Course
Study Period and Credits: 18 hours / 1 credit

课程介绍
　　本课程的主要任务是使建筑学/建筑技术学专业的学生掌握建筑环境的基本概念，学习建筑与城市热湿环境、风环境和空气质量的基础知识。通过课程教学，使学生熟悉城市微气候等理论，并了解人体对热湿环境的反应，掌握建筑环境学的实际应用和最新研究进展，为建筑能源和环境系统的测量与模拟打下坚实的基础。

Course Description
This course introduces students majoring in building science and engineering / building technology to the fundamentals of built environment. Students will study classical theories of built / urban thermal and humid environment, wind environment and air quality. Students will also familiarize urban micro environment and human reactions to thermal and humid environment. It is desirable that students will have the ability to measure and simulate building energy and environment systems based upon the knowledge of the latest development of the study of built environment.

研究生一年级
材料与建造·冯金龙
课程类型：必修
学时/学分：18学时/1学分

Graduate Program 1st Year
MATERIAL AND CONSTRUCTION · FENG Jinlong
Type: Required Course
Study Period and Credits: 18 hours / 1 credit

课程介绍
　　本课程介绍现代建筑技术的发展过程，论述现代建筑技术及其美学观念对建筑设计的重要作用。探讨由材料、结构和构造方式所形成的建筑建造的逻辑方式研究。研究建筑形式产生的物质技术基础，诠释现代建筑的建构理论与研究方法。

Course Description
It introduces the development process of modern architecture technology and discusses the important role played by the modern architecture technology and its aesthetic concepts in the architectural design. It explores the logical methods of construction of the architecture formed by materials, structure and construction. It studies the material and technical basis for the creation of architectural form, and interprets construction theory and research methods for modern architectures.

研究生一年级
计算机辅助技术·吉国华
课程类型：选修
学时/学分：36学时/2学分

Graduate Program 1st Year
TECHNOLOGY OF CAAD · JI Guohua
Type: Elective Course
Study Period and Credits: 36 hours / 2 credit

课程介绍
　　随着计算机辅助建筑设计技术的快速发展，当前数字技术在建筑设计中的角色逐渐从辅助绘图转向了真正的辅助设计，并引发了设计的革命和建筑的形式创新。本课程讲授AutoCAD VBA和RhinoScript编程。让学生在掌握"宏"/"脚本"编程的同时，增强以理性的过程思维方式分析和解决设计问题的能力，为数字建筑设计打下必要的基础。
　　课程分为三个部分：
　　1. VB语言基础，包括VB基本语法、结构化程序、数组、过程等编程知识和技巧；
　　2. AutoCAD VBA，包括AutoCAD VBA的结构、二维图形、人机交互、三维对象等，以及基本的图形学知识；
　　3. RhinoScript概要，包括基本概念、Nurbs概念、VBScript简介、曲线对象、曲面对象等。

Course Description
Following its fast development, the role of digital technology in architecture is changing from computer-aided drawing to real computer-aided design, leading to a revolution of design and the innovation of architectural form. Teaching the programming with AutoCAD VBA and RhinoScript, the lecture attempts to enhance the students' capability of reasoningly analyzing and solving design problems other than the skills of "macro" or "script" programming, to let them lay the base of digital architectural design.
The course consists of three parts:
1. Introduction to VB, including the basic grammar of VB, structural program, array, process, etc.
2. AutoCAD VBA, including the structure of AutoCAD VBA, 2D graphics, interactive methods, 3D objects, and some basic knowledge of computer graphics.
3. Brief introduction of RhinoScript, including basic concepts, the concept of Nurbs, sammary of VBScript, and Rhino objects.

研究生一年级
GIS基础与应用·童滋雨
课程类型：选修
学时/学分：18学时/1学分

Graduate Program 1st Year
CONCEPT AND APPLICATION OF GIS · TONG Ziyu
Type: Elective Course
Study Period and Credits: 18 hours / 1 credit

课程介绍
　　本课程的主要目的是让学生理解GIS的相关概念以及GIS对城市研究的意义，并能够利用GIS软件对城市进行分析和研究。

Course Description
This course aims to enable students to understand the related concepts of GIS and the significance of GIS to urban research, and to be able to use GIS software to carry out urban analysis and research.

其他
MISCELLANEA

# 讲座 Lectures

SOCIAL CARTOGRAPHY, ARTICULATORY URBANISM and PUBLIC ARCHITECTURE
社会制图、节点式城市建筑与公共建筑

讲座人
Felipe Hernandez

The VENTILATION of the BRITISH HOUSES of PARLIAMENT and the 19TH-CENTURY EXPERIMENTAL TRADITION
英国议会大厦的通风设计与19世纪建筑中的实验传统

讲座人
Henrik Schoenefeldt

THE 3RD CNRCAU FORUM ON ARCHITECTURAL THINKING

THEME OF THE FORUM
Environmental Architectonics
环境的建构

PARTICIPANTS

旧城改建中的国家主导
——以上海的城中村为例

吴缚龙
英国 UCL 教授
南京大学兼职教授

建筑　　　数学

秦佑国（清华大学建筑学院 教授 博导）

近代中国建筑学的黎明
——由日本传递的西方建筑文化

徐苏斌
天津大学 建筑学院
教授

南京大学鼓楼校区教学楼205室
二零一三年十一月十一日星期一上午十点十分

城市与影像：小屏幕与大屏幕的演进
Urbanity and Image: Micro and Macro Screen Evolutions

Richard Koeck
Professor
School of Architecture, Liverpool University
利物浦大学建筑系学院教授

南京大学蒙民伟楼十楼大教室
二零一三年十一月十四日星期四晚上七点整

消失的侗族建筑
Dong People's Vanishing Architecture

Klaus Zweger
克劳斯·茨威格

Faculty of Architecture and Planning, the Vienna University of Technology
维也纳理工大学建筑与规划学院
Associate professor
副教授

南京大学蒙民伟楼十楼大教室
二零一三年十一月十五日星期五晚七点整

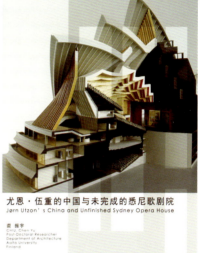

尤恩·伍重的中国与未完成的悉尼歌剧院
Jørn Utzon's China and Unfinished Sydney Opera House

袁振宇
Chiu, Chen Yu
Post-Doctoral Researcher
Department of Architecture
Aalto University
Finland

南京大学鼓楼校区蒙民伟楼1003教室
二零一四年三月七日星期五晚七点

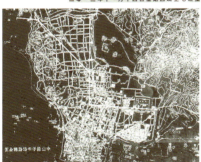

中国近代城市规划的演进

李百浩
东南大学 教授 博导

南京大学鼓楼校区教学楼205室
二零一三年十一月二十五日星期一上午十点十分

小说一点儿北京

李涵

南京大学蒙民伟楼十楼大教室
二零一三年十二月二十七日星期五晚六点整

美国的新城市主义——历史、现状与未来
New Urbanism in United States – Past, Present and Future

DENIS HECTOR
Associate Professor and Acting Dean
副教授、代院长
University of Miami School of Architecture
迈阿密大学建筑学院

南京大学蒙民伟楼1003教室
二零一四年三月十一日星期二下午五点整

# 硕士学位论文列表
# List of Thesis for Master Degree

| 研究生姓名 | 研究生论文标题 | 导师姓名 |
|---|---|---|
| 曹永山 | 深圳市前海企业馆及公建项目建筑方案设计 | 张 雷 |
| 耿 健 | 重庆黎香湖北岸一期工程方案设计 | 张 雷 |
| 杭晓萌 | 南京河西滨江万景园段小教堂方案及工程设计 | 张 雷 |
| 贾福有 | 南京江心洲红星教堂复建项目建筑设计 | 张 雷 |
| 王 彬 | 报恩寺遗址公园黄泥塘配套服务区规划及建筑设计 | 张 雷 |
| 陈焕彦 | 南京建邺区南湖社区组合服务中心设计研究 | 冯金龙 |
| 陈中高 | 南京名家科技大厦地块项目规划建筑方案设计 | 冯金龙 |
| 范丹丹 | 江苏文交所广场爱涛路美术馆设计 | 冯金龙 |
| 黄一庭 | 南京雨花经济区板桥A地块中小学设计研究 | 冯金龙 |
| 陈 成 | 准周期平面镶嵌在基于日照分析的建筑表皮形式设计中的应用 | 吉国华 |
| 司秉卉 | 基于日照分析的表皮形式设计 | 吉国华 |
| 余 露 | 基于可展开面的高层建筑造型研究 | 吉国华 |
| 陈 娟 | 资金（溧水）科创中心建筑设计及构造研究 | 周 凌 |
| 徐怡雯 | 莫干山南路乡乡公所改造与新建设计 | 周 凌 |
| 林中格 | 南京市溧水县石湫影视基地度假酒店设计 | 傅 筱 |
| 邵一丹 | 麒麟科技园七号地块食堂改造设计 | 傅 筱 |
| 潘 东 | 旧建筑改造更新设计研究——以芳桥供销社改造设计为例 | 丁沃沃 |
| 杨 灿 | 旧城更新中肌理织补方法设计研究——以芳桥镇老街肌理织补为例 | 丁沃沃 |
| 郑国活 | 传统街巷空间改造设计研究——以芳桥老街改造为例 | 丁沃沃 |
| 柯国新 | 基于形式语法的城市形态表述方法初探——以北京、南京传统城市街区为例 | 丁沃沃 |

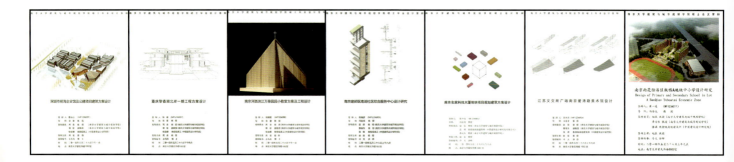

| 研究生姓名 | 研究生论文标题 | 导师姓名 |
|---|---|---|
| 陶敏悦 | 南京河西城市生态公园生态展示馆设计研究 | 华晓宁 |
| 王 凯 | 基于自然通风和热环境优化的新农居建筑设计 | 华晓宁 |
| 黄文华 | 闽北传统建造体系的现代化应用研究——以政和杨源水尾区域民宿设计为例 | 赵 辰 |
| 陆 恬 | 废弃水泥厂中高层厂房的改造利用设计研究——以慈城水泥厂改造为例 | 赵 辰 |
| 岳文博 | 废弃水泥厂改造中空间竖向高程关系的适应性研究——以慈城水泥厂改造为例 | 赵 辰 |
| 沈周娅 | 图解静力学结合参数化在自由形式设计中的应用 | 赵 辰 |
| 胡鹭鹭 | 青岛即墨传统院宅空间类型及建造特征浅析 | 萧红颜 |
| 辛胤庆 | 中国现代建筑个案研究——以同济大学教工俱乐部为例 | 王骏阳 |
| 胡 昊 | 南京中华路（古御道）研究——从东吴到现在 | 胡 恒 |
| 陈 园 | 面向夏热冬冷地区湿热气候的"绿建设计工作室VDS"本地化研究 | 秦孟昊 |
| 胡绮玭 | 宜兴芳桥镇蚕种场改造设计 | 鲁安东 |

在校学生名单
List of Students

## 本科生 Undergraduate

### 2010级学生 / Students 2010

| 陈博宇 CHEN Boyu | 胡任元 HU Renyuan | 刘树豪 LIU Shuhao | 谭 健 TAN Jian | 杨天仪 YANG Tianyi |
| --- | --- | --- | --- | --- |
| 陈凌杰 CHEN Lingjie | 黄广伟 HUANG Guangwei | 刘思彤 LIU Sitong | 王 琳 WANG Lin | 杨玉菡 YANG Yuhan |
| 陈晓敏 CHEN Xiaomin | 蒋 婷 JIANG Ting | 刘文沛 LIU Wenpei | 夏候蓉 XIA Hourong | 姚 梦 YAO Meng |
| 陈修远 CHEN Xiuyuan | 李乐之 LI Lezhi | 刘 宇 LIU Yu | 徐 晏 XU Yan | 张明杰 ZHANG Mingjie |
| 程 斌 CHENG Bin | 李平乐 LI Pingle | 鲁光耀 LU Guangyao | 许梦逸 XU Mengyi | 周平浪 ZHOU Pinglang |
| 顾一蝶 GU Yidie | 林 治 LIN Zhi | | | |

### 2011级学生 / Students 2011

| 崔傲寒 CUI Aohan | 蒋建昕 JIANG Jianxin | 彭丹丹 PENG Dandan | 王新宇 WANG Xinyu | 张豪杰 ZHANG Haojie |
| --- | --- | --- | --- | --- |
| 冯 琪 FENG Qi | 蒋造时 JIANG Zaoshi | 宋富敏 SONG Fumin | 吴家禾 WU Jiahe | 张黎萌 ZHANG Limeng |
| 顾聿笙 GU Yusheng | 雷朝荣 LEI Zhaorong | 拓 展 TUO Zhan | 吴结松 WU Jiesong | 张人祝 ZHANG Renzhu |
| 黄凯峰 HUANG Kaifeng | 黎乐源 LI Leyuan | 王梦琴 WANG Mengqin | 席 弘 XI Hong | 周 松 ZHOU Song |
| 黄雯倩 HUANG Wenqian | 柳纬宇 LIU Weiyu | 王却奁 WANG Quelian | 谢忠雄 XIE Zhongxiong | 周贤春 ZHOU Xianchun |
| 贾福龙 JIA Fulong | 缪姣姣 MIAO Jiaojiao | 王思绮 WANG Siqi | 徐亦杨 XU Yiyang | 左 思 ZUO Si |
| 蒋佳瑶 JIANG Jiayao | 倪若宁 NI Ruoning | 王潇聆 WANG Xiaoling | 杨益晖 YANG Yihui | |

### 2012级学生 / Students 2012

| 陈虹全 CHEN Hongquan | 葛嘉许 GE Jiaxu | 刘贤斌 LIU Xianbin | 沈应浩 SHEN Yinghao | 臧 倩 ZANG Qian | 赵媛倩 ZHAO Yuanqian |
| --- | --- | --- | --- | --- | --- |
| 陈思涵 CHEN Sihan | 桂 喻 GUI Yu | 刘姿佑 LIU Ziyou | 苏 彤 SU Tong | 张馨元 ZHANG Xinyuan | |
| 陈 妍 CHEN Yan | 黄福运 HUANG Fuyun | 陆怡人 LU Yiren | 唐林松 TANG Linsong | 张逸凡 ZHANG Yifan | |
| 从 彬 CONG Bin | 黄卫健 HUANG Weijian | 罗 坤 LUO Kun | 王 焱 WANG Yan | 朱朝龙 ZHU Chaolong | |
| 段晓昱 DUAN Xiaoyu | 黄子恩 HUANG Zi'en | 钱宇飞 QIAN Yufei | 王一侬 WANG Yinong | 朱凌峥 ZHU Lingzheng | |
| 高文杰 GAO Wenjie | 季惠敏 JI Huimin | 钱雨翀 QIAN Yuchong | 吴峥嵘 WU Zhengrong | 田 甜 TIAN Tian | |
| 高祥震 GAO Xiangzhen | 李慧兰 LI Huilan | 全道薰 QUAN Daoxun | 于明霞 YU Mingxia | 徐 华 XU Hua | |

### 2013级学生 / Students 2013

| 曹舒琪 CAO Shuqi | 黄婉莹 HUANG Wanying | 罗晓东 LUO Xiaodong | 王 青 WANG Qing | 徐家炜 XU Jiawei | 赵中石 ZHAO Zhongshi |
| --- | --- | --- | --- | --- | --- |
| 陈 露 CHEN Lu | 黄追日 HUANG Zhuiri | 吕 童 LU Tong | 王秋锐 WANG Qiurui | 徐瑜灵 XU Yuling | 周 怡 ZHOU Yi |
| 董素宏 DONG Suhong | 吉雨馨 JI Yuxin | 楠田康雄 KUSUDA YASUO | 王 瑶 WANG Yao | 杨 蕾 YANG Lei | |
| 郭金未 GUO Jinwei | 贾奕超 JIA Yichao | 宋宇瑋 SONG Yuxun | 王智伟 WANG Zhiwei | 元大海 YUAN Dahai | |
| 郭 硕 GUO Shuo | 林之音 LIN Zhiyin | 谭 皓 TAN Hao | 武 波 WU Bo | 章太雷 ZHANG Tailei | |
| 贺唯嘉 HE Weijia | 刘稷祺 LIU Jiqi | 涂成祥 TU Chengxiang | 夏凡琦 XIA Fanqi | 赵 焦 ZHAO Jiao | |
| 胡慧慧 HU Huihui | 鲁 晴 LU Qing | 王成阳 WANG Chengyang | 夏 楠 XIA Nan | 赵梦娣 ZHAO Mengdi | |

## 研究生 Postgraduate

鲍丽丽 BAO Lili  
郭 芳 GUO Fang  
李苑常 LI Yuanchang  
金筱敏 JIN Xiaomin  
王 一 WANG Yi  
俞 英 YU Ying  

谌 利 CHEN Li  
黄志鹏 HUANG Zhipeng  
林天予 LIN Tianyu  
汤梦捷 TANG Mengjie  
闻金石 WEN Jinshi  
曾宇城 ZENG Yucheng  

程 璐 CHENG Lu  
李日影 LI Riying  
刘 昀 LIU Jun  
涂梦如 TU Mengru  
吴 宁 WU Ning  
赵 慧 ZHAO Hui  

丁文博 DING Wenbo  
李善超 Li Shanchao  
柳 楠 LIU Nan  
王鑫星 WANG Xinxing  
吴仕佳 WU Shijia  
朱俊杰 ZHU Junjie  

丁文磊 DING Wenlei  
李莹莹 LI Yingying  
罗思维 LUO Siwei  
王雅谦 WANG Yaqian  
吴 玺 WU Xi  
朱 珠 ZHU Zhu  

---

陈 肯 CHEN Ken  
柯国新 KE Guoxin  
马 喆 MA Zhe  
吴绉彦 WU Zhouyan  
袁 芳 YUAN Fang  
葛鹏飞 GE Pengfei  
倪力均 NI Lijun  
王 晨 WANG Chen  
张 岸 ZHANG An  

陈 圆 CHEN Yuan  
黎健波 LI Jianbo  
孟庆忠 MENG Qingzhong  
谢智峰 XIE Zhifeng  
张 备 ZHANG Bei  
胡 曜 HU Yao  
潘 旻 PAN Min  
夏 澍 XIA Shu  
张 敏 ZHANG Min  

陈 钊 CHEN Zhao  
李 港 LI Gang  
沈周娅 SHEN Zhouya  
辛胤庆 XIN Yinqing  
张卜予 ZHANG Buyu  
金 鑫 JIN Xin  
彭文楷 PENG Wenkai  
徐庆姝 XU Qingshu  
张 培 ZHANG Pei  

高 菲 GAO Fei  
李恒鑫 LI Hengxin  
石延安 SHI Yan'an  
徐 睿 XU Rui  
曹梦原 CAO Mengyuan  
李红瑞 LI Hongrui  
乔 力 QIAO Li  
姚丛琦 YAO Congqi  
张永磊 ZHANG Yonglei  

管 理 GUAN Li  
刘滨洋 LIU Binyang  
王海芹 WANG Haiqin  
杨尚宜 YANG Shangyi  
陈 姝 CHEN Shu  
李 扬 LI Yang  
邱金宏 QIU Jinhong  
于海平 YU Haiping  
赵 锐 ZHAO Rui  

韩 梦 HAN Meng  
刘兴渝 LIU Xingyu  
王力凯 WANG Likai  
殷 奕 YIN Yi  
陈婷婷 CHEN Tingting  
刘 宇 LIU Yu  
沈均臣 SHEN Junchen  
虞王璐 YU Wanglu  
赵天亚 ZHAO Tianya  

胡 昊 HU Hao  
刘奕彪 LIU Yibiao  
王亦播 WANG Yibo  
郁新新 YU Xinxin  
陈 新 CHEN Xin  
吕 程 Lu Cheng  
汪 园 WANG Yuan  
袁金燕 YUAN Jinyan  
周 逸 ZHOU Yi  

胡鹭鹭 HU Lulu  
吕 铭 Lu Ming  

---

曹永山 CAO Yongshan  
樊璐敏 FAN Lumin  
黄文华 HUANG Wenhua  
林中格 LIN Zhongge  
司秉卉 SI Binghui  
吴黎明 WU Liming  
杨 浩 YANG Hao  
袁亮亮 YUAN Liangliang  
赵书艺 ZHAO Shuyi  

陈 成 CHEN Cheng  
耿 健 GENG Jian  
黄一庭 HUANG Yiting  
刘赟俊 LIU Yunjun  
孙 燕 SUN Yan  
武苗苗 WU Miaomiao  
杨 骏 YANG Jun  
岳文博 YUE Wenbo  
赵潇欣 ZHAO Xiaoxin  

陈焕彦 CHEN Huanyan  
韩艺宽 HAN Yikuan  
贾福有 JIA Fuyou  
龙俊荣 LONG Junrong  
陶敏悦 TAO Minyue  
徐怡雯 XU Yiwen  
杨 柯 YANG Ke  
张 成 ZHANG Cheng  
郑国活 ZHENG Guohuo  

陈 娟 CHEN Juan  
杭晓萌 HANG Xiaomeng  
蒋菁菁 JIANG Jingjing  
陆 恬 LU Tian  
王 彬 WANG Bin  
薛晓旸 XUE Xiaoyang  
殷 强 YIN Qiang  
张方籍 ZHANG Fangji  
周 青 ZHOU Qing  

陈 鹏 CHEN Peng  
胡绮玭 HU Qipi  
赖友炜 LAI Youwei  
倪绍敏 NI Shaomin  
王洁琼 WANG Jieqiong  
颜骁程 YAN Xiaocheng  
余 露 YU Lu  
张 伟 ZHANG Wei  
周雨馨 ZHOU Yuxin  

陈中高 CHEN Zhonggao  
胡小敏 HU Xiaomin  
李 政 LI Zheng  
潘 东 PAN Dong  
王 凯 WANG Kai  
杨 灿 YANG Can  
俞 冰 YU Bing  
张文婷 ZHANG Wenting  
朱鹏飞 ZHU Pengfei  

范丹丹 FAN Dandan  
黄凯熙 HUANG Kaixi  
林肖寅 LIN Xiaoyin  
邵一丹 SHAO Yidan  
王旭静 WANG Xujing  
杨钗芳 YANG Chaifang  
俞 琳 YU Lin  
赵 芹 ZHAO Qin  
朱 煜 ZHU Yu  

---

奥珅颖 AO Shenying  
雷冬雪 LEI Dongxue  
孟文儒 MEHG Wenru  
王珊珊 WANG Shanshan  
许伯晗 XU Bohan  
郭 瑛 GUO Ying  
力振球 LI Zhenqiu  
孙 昕 SUN Xin  
徐 蕾 XU Lei  

陈观兴 CHEN Guanxing  
李 彤 LI Tong  
潘柳青 PAN Liuqing  
魏江洋 WEI Jiangyang  
许 骏 XU Jun  
郭耘锦 GUO Yunjin  
刘 莹 LIU Ying  
谭发兵 TAN Fabing  
徐沁心 XU Qinxin  

陈相莒 CHEN Xiangying  
李招成 LI Zhaocheng  
潘幼健 PAN Youjian  
吴超楠 WU Chaonan  
曹 政 CAO Zheng  
季 萍 JI Ping  
刘玉婧 LIU Yujing  
汤建华 TANG Jianhua  
徐婉迪 XU Wandi  

段艳文 DUAN Yanwen  
刘 佳 LIU Jia  
沙吉敏 SHA Jimin  
吴嘉鑫 WU Jiaxin  
陈 逸 CHEN Yi  
贾江南 JIA Jiangnan  
柳筱娴 LIU Xiaoxian  
王斌鹏 WANG Binpeng  
张 楠 ZHANG Nan  

符靓璇 FU Jingxuan  
刘彦辰 LIU Yanchen  
谭子龙 TAN Zilong  
夏 炎 XIA Yan  
仇高颖 QIU Gaoying  
姜 智 JIANG Zhi  
沈康惠 SHEN Kanghui  
王 晗 WANG Han  
赵 阳 ZHAO Yang  

黄龙辉 Huang Longhui  
吕 航 LU Hang  
王淡秋 WANG Danqiu  
肖 霄 XIAO Xiao  
戴 波 DAI Bo  
蒯冰清 KUAI Bingqing  
施 伟 SHI Wei  
王 倩 WANG Qian  
周荣楼 ZHOU Ronglou  

姜伟杰 JIANG Weijie  
毛军列 Mao Junlie  
王冬雪 WANG Dongxue  
徐少敏 XU Shaomin  
费日晓 FEI Rixiao  
李 昭 LI Zhao  
孙冠成 SUN Guancheng  
吴 宾 WU Bin

图书在版编目（CIP）数据

南京大学建筑与城市规划学院建筑系教学年鉴. 2013~2014 / 王丹丹，华晓宁编. -- 南京：东南大学出版社，2014.12
ISBN 978-7-5641-5353-3

Ⅰ. ①南… Ⅱ. ①王… ②华… Ⅲ. ①建筑学—教学研究—高等学校—南京市—2013~2014—年鉴②城市规划—教学研究—高等学校—南京市—2013~2014—年鉴 Ⅳ. ①TU-42

中国版本图书馆CIP数据核字（2014）第280940号

| | |
|---|---|
| 策　　划： | 丁沃沃　赵　辰 |
| 装帧设计： | 王丹丹　丁沃沃 |
| 责任编辑： | 姜　来　魏晓平 |
| 参与制作： | 颜骁程　陶敏悦 |

| | |
|---|---|
| 出版发行： | 东南大学出版社 |
| 社　　址： | 南京市四牌楼2号 |
| 出 版 人： | 江建中 |
| 网　　址： | http://www.seupress.com |
| 邮　　箱： | press@seupress.com |
| 邮　　编： | 210096 |
| 经　　销： | 全国各地新华书店 |
| 印　　刷： | 南京新世纪联盟印务有限公司 |
| 开　　本： | 787mm×1092mm　1/20 |
| 印　　张： | 8 |
| 字　　数： | 480千 |
| 版　　次： | 2014年12月第1版 |
| 印　　次： | 2014年12月第1次印刷 |
| 书　　号： | ISBN 978-7-5641-5353-3 |
| 定　　价： | 58.00元 |

本社图书若有印装质量问题，请直接与营销部联系。电话：025-83791830